The Coccidian Parasites
(Protozoa, Apicomplexa) of Carnivores

The Coccidian Parasites
(Protozoa, Apicomplexa) of
Carnivores

NORMAN D. LEVINE and VIRGINIA IVENS

ILLINOIS BIOLOGICAL MONOGRAPHS 51

UNIVERSITY OF ILLINOIS PRESS Urbana Chicago London

ILLINOIS BIOLOGICAL MONOGRAPHS

Volumes 1 through 24 contained four issues each. Beginning with number 25 (issued in 1957), each publication is numbered consecutively. Standing orders are accepted for forthcoming numbers. The titles listed below are still in print. They may be purchased from the University of Illinois Press, 54 East Gregory Drive, Box 5081, Station A, Champaign, Illinois 61820. Out-of-print titles in the Illinois Biological Monographs are available from University Microfilms, Inc., 300 North Zeeb Road, Ann Arbor, Michigan 48106.

KOCH, STEPHEN D. (1974): The *Eragrostis-pectinacea-pilosa* Complex in North and Central America (Gramineae: Eragrostoideae). 86 pp. 14 figs. 8 plates. No. 48. $5.95.

KENDEIGH, CHARLES S. (1979): Invertebrate Populations of the Deciduous Forest: Fluctuations and Relations to Weather. 153 pp. Illus. Tables. No. 50. $10.00.

Library of Congress Cataloging in Publication Data

Levine, Norman D
    The coccidian parasites (Protozoa, Apicomplexa)
of carnivores.

    (Illinois biological monographs; 51)
    Bibliography: p.   82-8649
    Includes index.
    1. Coccidia.   2. Parasites—Carnivora.
3. Carnivora—Diseases.   I. Ivens, Virginia,
joint author.   II. Title.   III. Series.   [DNLM:
1. Carnivora.   2. Protozoa.   3. Parasites.   W1
IL237 no. 51/QL 737.C2 L665c]
QL368.C7L57   599.05'249   80-25205
ISBN 0-252-00856-1

Contents

Introduction

According to Walker et al. (1975), the mammalian order Carnivora contains 7 families, 101 genera, and about 248 species. Most species in the order are predators, but some are not.

This is a fairly small group, yet the problems of coccidian parasitism that it presents are considerable. Only a few years ago it was thought that all that remained was to examine the previously unexamined host species and describe their coccidia, which should be little different from those already known. But the situation has changed completely. A few years ago the only genera of coccidia thought to occur in the group were *Eimeria, Isospora,* and *Cryptosporidium.* Now we know that *Toxoplasma, Sarcocystis, Besnoitia,* and *Frenkelia* are all coccidia with oocysts similar to those of *Isospora* and that most of the organisms given the name *Cryptosporidium* are actually *Sarcocystis.* We also know that all coccidia are not homoxenous (with a single host in the life cycle) as we had supposed, but that some of them are heteroxenous, with one part of their life cycle in one host and another part in another host. The host of the sexual stages is a predator, and that of the asexual stages is a prey animal; the former becomes infected by eating the latter, or at least its tissues.

Before discussing the various coccidian species, it is necessary to describe their life cycle in broad outline, and to give diagnoses

This monograph is a contribution from the College of Veterinary Medicine and the Agricultural Experiment Station, University of Illinois. It was supported in part by National Science Foundation grants no. GB-30800X and no. BMS71-02194.

for those genera that occur in the Carnivora. Other genera, which have not been found in this host order, are not mentioned herein. Coccidia *sensu stricto* belong to the protozoan phylum Apicomplexa Levine, 1970; class Sporozoasida Leuckart, 1879; subclass Coccidiasina Leuckart, 1879; order Eucoccidiorida Léger and Duboscq, 1910; suborder Eimeriorina Léger, 1911. Within this suborder they are divided into 11 families, but the only ones that concern us here are the family Eimeriidae Minchin, 1903 and the family Sarcocystidae Poche, 1913. Within the Eimeriidae are 13 genera; within the Sarcocystidae, there are 5. The diagnoses of those genera that occur in carnivores are:

Phylum APICOMPLEXA Levine, 1970
Apical complex, generally consisting of polar ring(s), rhoptries, micronemes, conoid, and subpellicular microtubules present at some stage; micropore(s) generally present at some stage; cilia absent; sexuality by syngamy; all species parasitic.
Class SPOROZOASIDA Leuckart, 1879
If present, conoid forms complete cone; reproduction generally both sexual and asexual; oocysts contain infective sporozoites that result from sporogony; locomotion by body flexion, gliding, undulation of longitudinal ridges or flagellar lashing; flagella present only in microgametes of some groups; pseudopods ordinarily absent, if present used for feeding, not locomotion; homoxenous or heteroxenous.
Subclass COCCIDIASINA Leuckart, 1879
Gamonts ordinarily present; mature gamonts small, typically intracellular; conoid not modified into mucron or epimerite; syzygy generally absent, if present involves gametes; anisogamy marked; life cycle characteristically consists of merogony, gametogony, and sporogony; most species in vertebrates.
Order EUCOCCIDIORIDA Léger and Duboscq, 1910
Merogony present; in vertebrates and/or invertebrates.
Suborder EIMERIORINA Léger, 1911
Macrogamete and microgamont develop independently; syzygy absent; microgamont typically produces many microgametes; zygote not motile; sporozoites typically enclosed in a sporocyst; endodyogeny absent or present; homoxenous or heteroxenous.

Family EIMERIIDAE Minchin, 1903
Development in host cell proper; without attachment organelle or "vaginal" tube; oocysts with 0, 1, 2, 4, or more sporocysts, each with 1 or more sporozoites; homoxenous or at least without asexual multiplication in nondefinitive host; merogony within host, sporogony typically outside; metrocytes not formed; microgametes with 2 or 3 flagella; in vertebrates or invertebrates.
Genus *Eimeria* Schneider, 1875
Oocysts with 4 sporocysts, each with 2 sporozoites; about 1,030 named species in vertebrates and invertebrates.
Genus *Isospora* Schneider, 1881
Oocysts with 2 sporocysts, each with 4 sporozoites; about 200 named species in vertebrates and invertebrates.
Family SARCOCYSTIDAE Poche, 1913
Heteroxenous, producing oocysts following syngamy; development in host cell proper; without attachment organelle or "vaginal" tube; oocysts with 2 sporocysts, each with 4 sporozoites, in intestine of a definitive host; with asexual stages in an intermediate host.
Subfamily SARCOCYSTINAE Poche, 1913
Obligatorily heteroxenous; asexual multiplication in intermediate (prey) host; last generation meronts ("sarcocysts") in intermediate host form metrocytes, which give rise to bradyzoites, which are infectious for definitive (predator) host; oocysts sporulate in predator host tissues; sporulated sporocysts in feces.
Genus *Sarcocystis* Lankester, 1882
Last generation meronts typically in striated muscles; merozoites elongate; about 88 named species.
Genus *Frenkelia* Biocca, 1968
Last generation meronts typically in central nervous system; merozoites elongate; 2 named species.
Subfamily TOXOPLASMATINAE Biocca, 1957
Complete life cycle obligatorily heteroxenous except in the Felidae, but asexual stages usually transmissible from one intermediate host to another; metrocytes

not formed; oocysts do not sporulate in host tissues.
Genus *Toxoplasma* Nicolle and Manceaux, 1909
Meronts in many types of cell; host cell nucleus
outside meront wall; 7 named species.
Genus *Besnoitia* Henry, 1913
Meronts in fibroblasts and probably other cells; host
cell nuclei within meront wall; 7 named species.

No new genera are introduced, but the synonymy of the various species is corrected. Some of the synonyms are quite old and well accepted, but a few were invented by some modern workers who ignored the International Rules of Zoological Nomenclature and introduced new names of their own for some species of hetero-xenous coccidia (see Levine, 1977a).

It is possible that future research will necessitate some changes in the above diagnoses.

A typical oocyst is that of *Eimeria* (Fig. 1). Apicomplexan fine structures are shown in Fig. 2. These structures are present in merozoites. Sporozoites are similar in appearance to merozoites, but they usually have a clear (refractile) globule composed of proteinaceous material plus apparently amylopectin at the broad end, and sometimes a similar one near the small end, whereas merozoites do not. Metrocytes are similar to merozoites, but are broader and have a deeply folded cell surface with more than 1 micropore, rudimentary rhoptries, and relatively few micronemes.

A typical coccidian life cycle is shown in Fig. 3. The oocysts are ingested by a host animal, and the sporozoites are released from them. The sporozoites enter host cells, turn into meronts, and multiply by merogony, forming a variable number of mero-zoites by multiple fission. These enter new host cells, turn into meronts, and again multiply by merogony. There is ordinarily a small, fixed number of asexual generations. The last generation merozoites enter new host cells and become macrogametes, which simply grow, or become microgamonts, which multiply asexually by multiple fission to form a large number of flagellated micro-gametes. Syngamy takes place to form a zygote; a wall is laid down around it, and it becomes an oocyst. The oocyst ordinarily passes out of the host and does not develop further (sporulate) until it reaches the outside. However, sporulation takes place within the host's body in *Sarcocystis* and *Frenkelia* (and also in some species of *Eimeria* and *Isospora* in fish and reptiles). In sporulation, secondary cysts (sporocysts) develop within the oocysts, and sporozoites (infective cells) develop within them.

In some species of *Isospora*, a sporozoite may enter a lymph node or other cell of a foreign (transport) host animal in which it cannot mature and grow into a giant sporozoite (hypnozoite). It can then infect its natural host when the latter eats the transport host.

In *Sarcocystis*, the asexual part of the life cycle takes place in a prey host, and the sexual part in a predator host. We know the complete life cycles of only a few species so far, but what seems to occur is that 1 or 2 generations of tachyzoite occur in the endothelial cells of the small blood vessels of various organs and tissues, and these then form a final generation of bradyzoites in sarcocysts in the muscle cells. Only the bradyzoites are infective for the predator host, and it obtains them by eating the prey host or its infected muscle. The bradyzoites turn into gamonts in the intestinal cells of the predator. The gamonts form male and female gametes, these fuse to form zygotes, and the zygotes develop into mature oocysts; these develop into mature sporocysts that pass out in the feces and are infective for the prey host.

The life cycle of the Toxoplasmatinae differs from that of *Sarcocystis* in that asexual development can take place in both prey and predator hosts (the latter always being Felidae so far as is known) and in that the oocysts passed in the feces of the predator host are immature and sporulation then takes place on the ground. Further, both tachyzoites and bradyzoites are capable of infecting new intermediate hosts; the tachyzoites occur in many types of cells, and especially in the peritoneal cavity, while the bradyzoites are found in pseudocysts, especially in the brain. More details about the life cycles of individual Sarcocystidae are given below.

There are 3 schizogonies—merogony, gametogony, and sporogony. Coccidia are haploid throughout their life cycle except in the zygote stage. The first sporogonous division is meiotic, reestablishing the haploid condition.

Systematic Section

The sequence and names of carnivore families and genera used below are those of Walker et al. (1975), except that *Vulpes fulva* is accepted as a synonym of *V. vulpes* (R. L. Rausch, personal communication). If some feature of a coccidian species is unknown (and many are), it has been omitted in the interest of economy.

Host Family CANIDAE

Host Genus *Canis*

Eimeria canis Wenyon, 1923

(Plate 1, Figs. 3, 4, 5)

Type Host. Domestic dog *Canis familiaris.*
Other Hosts. Dingo *C. dingo,* coyote *C. latrans,* domestic cat *Felis catus* (?).
Location. Unknown; oocysts found in feces.
Geographic Distribution. Worldwide.
Prevalence. Uncommon.
Oocyst Structure. Ovoid or ellipsoidal, 17–45 x 11–28 μm, with fairly thick, rough, 2-layered, colorless, pink or red wall, with micropyle. Sporocysts 7–12 x 7–8 (mean, 9 x 7) μm. Sporozoites 9–11 x 2.5 (mean, 9.5 x 2.5) μm.
Sporulation. 1–4 days.
Remarks. It is far from certain that this is a valid species. Wenyon (1926) remarked that in many respects it resembled a mixture of *E. stiedai* and *E. perforans* of the rabbit; Goodrich (1944) thought

it was a rabbit form that dogs had eaten, and there is nothing in later descriptions to contradict the possibility that it is a rabbit form. Dubey, Fayer, and Seesee (1978) found *Eimeria* sp. oocysts in the feces of 9% of 169 adult coyotes *C. latrans* in Montana; they probably originated in prey animals that the coyotes had eaten.

Eimeria rayii Rao and Bhatavdekar, 1957

Type Host. Domestic dog *Canis familiaris.*
Location. Unknown; oocysts found in feces.
Geographic Distribution. Asia (India).
Prevalence. Unknown.
Oocyst Structure. Stout ovoid to slightly ellipsoidal, 22–29 x 18–22 (mean, 27 x 20) μm, with pale yellowish brown wall with a transparent outer part and a single heavy black contour line "between oocyst membrane and outer wall," with micropyle and micropylar cap. No other structural information given.
Remarks. This species was described from the feces of a dog in Bombay. Despite the authors' disclaimer, these oocysts resemble those of *E. ovina* of the sheep or *E. arloingi* of the goat, and were probably a pseudoparasite of the dog.

Isospora burrowsi Trayser and Todd, 1978

(Plate 3, Fig. 16)

Type Host. Domestic dog *Canis familiaris.*
Location. Posterior 3/5 of small intestine and cecum, above and below the host cell nuclei in epithelial cells and lamina propria cells of tips of villi.
Geographic Distribution. North America (New York).
Prevalence. Unknown.
Oocyst Structure. Sporulated oocysts spherical to ellipsoidal, 17–22 x 16–19 (mean, 20 x 17) μm, with smooth, yellow-green, 1-layered wall about 1 μm thick, without micropyle, residuum, or polar granule. Sporocysts ovoid to ellipsoidal, 12–16 x 8–11 (mean, 14 x 10) μm, without Stieda body, with large residuum.

Sporozoites elongate, with blunt posterior and somewhat tapering anterior end, about 6-8 x 4-5 (mean, 7 x 4) μm, not lying in any particular order in sporocysts.

Sporulation. Occurs outside host; time required not determined.

Merogony. According to Trayser (1973) and Trayser and Todd (1978), who used a cloned culture derived from a single oocyst, the sporozoites enter intestinal cells on the first day after inoculation and have developed into mature first-generation meronts on day 4. These are 11-18 x 9-18 (mean, 15 x 12) μm in tissue sections and are tightly packed with merozoites 8-14 x about 3-3.5 (mean, 11 x 3) μm. There was an average of 5 merozoites per meront of the tissue sections, but, of course, the number per meront was higher than that seen in a section. First-generation meronts were seen through day 6. Second-generation meronts were first seen 5 days after inoculation but were more numerous on day 6. They were 18-35 x 17-22 (mean, 26 x 18) μm in tissue sections, and contained rather loosely arranged, cylindrical merozoites with blunt ends that had a subcentral nucleus and measured 14-18 x 3-6 (mean, 16 x 5) μm.

Gametogony. Trayser (1973) and Trayser and Todd (1978) first saw immature microgamonts 6 days after inoculation. They enlarged, becoming 13-27 x 10-21 (mean, 20 x 14) μm, and containing a large number of microgametes 4-5 x 0.4 μm.

Immature macrogametes appeared 6 days after inoculation. They were elongate ovoid to spherical and 11-25 x 8-18 (mean, 17 x 11.5) μm when mature; they did not contain eosinophilic plastic granules (wall-forming bodies).

Prepatent Period. 6 days.

Patent Period. 11 days.

Isospora bahiensis de Moura Costa, 1956 emend. Levine, 1978

(Plate 2, Fig. 9)

Synonyms. Small *Isospora bigemina* (Stiles, 1891) Lühe, 1906 of *auctores; [non] Isospora bigemina* (Stiles, 1891) Lühe, 1906; *Isospora bigemina* Stiles, 1891 var. *bahiensis* de Moura Costa, 1956;

Isospora wallacei Dubey, 1976; *Isospora heydorni* Tadros and Laarman, 1976; *Hammondia heydorni* (Tadros and Laarman, 1976) Dubey, 1977.

Type Host. Domestic dog *Canis familiaris.*

Other Host. Probably red fox *Vulpes vulpes,* perhaps dingo *C. dingo.*

Location. Epithelial cells distal to host cell nucleus in tips of the villi throughout the small intestine (Dubey and Fayer, 1976).

Geographic Distribution. Worldwide. It is impossible to specify the localities where *I. bahiensis* has been found because it has probably been often reported merely as *I. bigemina,* a name that is now known to refer not only to *I. bahiensis* but also to several species of *Sarcocystis* (see Levine, 1977a). However, it is known that *I. bahiensis* has been found in dogs in England (Wenyon and Sheather, 1925; Wenyon, 1926), Germany (Heydorn, 1973), the Netherlands (Tadros and Laarman, 1976), Brazil (de Moura Costa, 1956), Illinois (Levine and Ivens, 1965a), Maryland (Fayer, 1974), Colorado (Gassner, 1940), Iowa (Lee, 1934), probably in the red fox *V. vulpes* in Bulgaria (Golemanski and Ridzhakov, 1975) and perhaps in the dingo *C. dingo* in the Leningrad zoo (Yakimoff and Matchulski, 1935).

Prevalence. Apparently not common. Gassner (1940) found what he called the small form of *I. bigemina* in 71% of 320 dogs in Colorado; perhaps it was *I. bahiensis.* However, it has not been found there in recent years (W. C. Marquardt, personal communication). Levine and Ivens (1965a) found it in 1 of 139 dogs in Illinois. Because this species has been confused with *Sarcocystis bigemina* in the past, other reports on prevalence cannot be relied on.

Oocyst Structure. The following description is from Levine and Ivens (1965a). Unsporulated oocysts subspherical, occasionally spherical or broadly ellipsoidal, very pale, 10–14 x 10–12 (mean, 12 x 10) μm, with smooth, colorless, 1-layered wall about 0.4-μm thick. Sporulated oocysts subspherical to broadly ellipsoidal, very pale, 12–14 x 10–12 (mean, 13 x 11) μm, with smooth, colorless, 1-layered wall 0.4-μm thick, without micropyle or residuum, with polar granule present initially but disappearing after sporulation and storage for 6 days. Sporocysts broadly ellipsoidal, 7–8 x 5–7 (mean, 8 x 6) μm, without Stieda body, with residuum. Sporozoites sausage-shaped, generally all oriented in same direction within sporocyst, uniformly pale, without clear globule.

Sporulation. 12 hours at 30–37 C, 1–3 days at room temperature (de Moura Costa, 1956; Heydorn, 1973; Dubey and Fayer, 1976).

Merogony. The endogenous stages were described by Dubey and Fayer (1976). They found meronts 5–7 μm in diameter in epithelial cells at the tips of the small intestine villi. These meronts contained 3–12 merozoites 5–6 x 1–2 μm, without a residuum.

Gametogony. Dubey and Fayer (1976) found gamonts and gametes in the same location as meronts. The microgamonts were 5–8 μm in diameter and contained 6–12 microgametes. The macrogametes were 7–10 μm in diameter.

Prepatent Period. 7–15 days after ingestion of oocysts (Dubey and Fayer, 1976).

Patent Period. 1–9 days (Dubey and Fayer, 1976; Heydorn, 1973).

Pathogenicity. Apparently only slightly pathogenic. Dubey and Fayer (1976) found no evidence of pathogenicity, but the infected dog seen by Levine and Ivens (1965a) had a transient diarrhea.

Immunity. Heydorn (1973) said that dogs that had been previously infected with this species were not resistant to reinfection 6 or more weeks later. Heydorn (1973) found that the Sabin-Feldman dye test for *Toxoplasma* antibodies was negative with the sera of infected dogs.

Remarks. The nomenclature of this species has been confused for many years. The name *Coccidium bigeminum* was given by Stiles (1891) to a small coccidium that he found already sporulated in the dog intestine. It was placed in the genus *Isospora* by Lühe (1906). Wenyon and Sheather (1925) and Wenyon (1926) applied this name to 2 different small coccidia, one sporulated and the other not, which they found in the dog (see Levine, 1977a for details). In this erroneous action they were followed by many authors, and it became customary to speak of the small *I. bigemina* when referring to the unsporulated form and the large *I. bigemina* when referring to the sporulated one. The sporulated form was eventually transferred to the genus *Sarcocystis* as *S. bigemina*, leaving the unsporulated form presumably without a name. It was thereupon called *I. heydorni* by Tadros and Laarman (1976), *I.*

wallacei by Dubey (1976), and *Hammondia heydorni* (Tadros and Laarman, 1976) nov. comb. by Dubey (1977). However, de Moura Costa (1956) had already given it the name *I. bigemina* var. *bahiensis*, and this name has priority. Levine (1978), therefore, emended it to *I. bahiensis*.

There is still a good deal of doubt regarding the life cycle of this species. Dubey and Fayer (1976) were able to produce oocysts in the feces of dogs by feeding sporulated oocysts, but Heydorn (1973) was not. In other words, Dubey and Fayer (1976) reported a direct life cycle, while Heydorn (1973) reported an indirect one. Heydorn (1973) gave 12 conventional and 3 gnotobiotic beagle dogs several feedings of bovine esophagus; 9 conventional and all the gnotobiotic dogs passed oocysts of *I. bahiensis* for 1–9 days. He fed 2 calves 1 million oocysts each and killed them 8 and 12 weeks later, respectively. He then fed parts of these calves to dogs. The dogs fed liver of the 8-week calf did not pass oocysts, but those fed muscle (heart, diaphragm, esophagus, and skeletal) passed oocysts in their feces for 4–6 days after a delay of 9–10 days. The dogs fed diaphragm, esophagus, and skeletal muscle of the 12-week calf did not shed oocysts, but 2 dogs fed heart muscle passed *I. bahiensis* oocysts in their feces for 5–8 days after a delay of 7–10 days. He fed from 40 thousand to 1.3 million oocysts to 19 conventional and 2 gnotobiotic dogs and found no oocysts in their feces during 6 weeks of observation. On the other hand, he fed 9 dogs muscle tissue from dogs that had been fed *I. bahiensis* oocysts at least 7 weeks before and found that all of them passed oocysts in their feces for 2–9 days after a delay of 6–9 days.

In contrast, Dubey and Fayer (1976), in addition to the findings described above, found that dogs fed exposed cats or mice did not shed oocysts. They did find that 2 dogs fed hearts and diaphragms from naturally infected cattle shed oocysts 17 and 21 days after ingesting their first meal, but these oocysts were not infective for cattle. Nevertheless, they thought that a dog-ox-dog cycle was the natural mode of infection.

This is possibly the species that Golemanski and Ridzhakov (1975) found in the feces of 3% of 146 red foxes in Bulgaria. Its oocysts were unsporulated, 13–18 x 10–15 μm.

Isospora canis Nemeséri, 1959

(Plate 1, Fig. 1; Plate 2, Fig. 7)

Synonyms. *Diplospora bigemina* Wasielewsky, 1904 from dog
in part; *Isospora felis* (Wasielewsky, 1904) Wenyon, 1923 of *auctores* from the domestic dog; *Isospora bigemina* (Stiles, 1891)
Lühe, 1906 of *auctores* in part; *Levinea canis* (Nemeséri, 1959)
Dubey, 1977; *Cystoisospora canis* (Nemeséri, 1959) Frenkel, 1977.
Type Definitive Host. Domestic dog *Canis familiaris.*
Other Definitive Host. Coyote *C. latrans.*
Transport Hosts. House mouse *Mus musculus* (experimental),
newborn kittens *Felis catus* (experimental).
Location. Small intestine, mostly posterior third, and also to
some extent large intestine. Meronts mostly just beneath the epithelium, but a few deeper in the lamina propria; none occur in the
epithelial cells. Macrogametes and microgamonts in epithelial cells,
subepithelial connective tissue of villi of small intestine and also in
mucosa of large intestine (Nemeséri, 1960; Lepp and Todd, 1974).
Hypnozoites in lymph nodes of transport host.
Geographic Distribution. Worldwide.
Prevalence. Quite common. Gassner (1940) found it in 6% of 320
dogs in Colorado, Catcott (1946) in 3.5% of 113 in Ohio, Streitel
and Dubey (1976) in 2% of 500 stray dogs in Ohio; Levine and
Ivens (1965a) in 16% of 139 in Illinois, Lepp and Todd (1974) in
2% of 308 in Illinois, Choquette and Gelinas (1950) in 15% of 155
in Montreal, Nemeséri (1960) in 8% of 220 in Hungary, De Amaral,
Amaro, and Birgel (1964) in 1% of 232 in Brazil, Alcaino and Tagle
(1970) in 2% of 1,505 in Chile, and Mirza (1970) in 9% of 54 in
Iraq.
Dubey, Fayer, and Seesee (1978) found *I. canis*-like oocysts in
the feces of 2% of 169 adult coyotes *C. latrans* in Montana.
Oocyst Structure. Sporulated oocysts broadly ellipsoidal to
slightly ovoid, 34–42 x 27–33 (mean, 38 x 30) μm, with smooth,
very pale tan to light green, 1-layered wall 1.3–1.5 μm thick, sometimes appearing to be lined by a very thin membrane, without micropyle, residuum, or polar granule but with a tiny blob adherent
to the inside of the oocyst wall at the broad end. Sporocysts ellipsoidal, 18–24 x 15–18 (mean, 21 x 16) μm, with smooth, colorless

wall 0.4 μm thick, without Stieda body, with prominent residuum. Sporozoites sausage-shaped, usually oriented more or less lengthwise in sporocysts, with subcentral clear globule. Levine and Ivens (1965a) found a single *Caryospora*-like oocyst with a single sporocyst containing 8 sporozoites in 1 infected dog; they considered it to be an aberrant oocyst of *I. canis.*

The above description is from Neméséri (1959, 1960) and Levine and Ivens (1965a). Speer et al. (1973) found with the electron microscope that the sporocyst wall is composed of 2 layers, of which the outer is about 1/4 as thick as the inner. The latter is made up of 4 separate plates. Roberts, Mahrt, and Hammond (1972) described the fine structure of the sporozoites. They resemble those of *Eimeria* except that they have 2 crystalloid bodies in place of the clear globules found in *Eimeria,* and they have 26 subpellicular microtubules instead of the 22–25 of *Eimeria.* The particles in the anterior and posterior crystalloid bodies are irregularly arranged. The anterior one is closely associated with the rhoptries (number not stated) and the posterior one with amylopectin granules; endoplasmic reticulum and mitochondria are adjacent to both. There are many micronemes in the anterior region. There are a posterior pore, 2 anterior polar rings, and a conoid. Roberts, Mahrt, and Hammond (1972) thought that the crystalloid body and clear globule were similar in composition, perhaps associated with food storage, and not viral in nature. They saw 2 micropores in 1 sporozoite, but only 1 in others.

Sporulation. Described by Lepp and Todd (1976). It takes 48 hours at 20 C, 16 hours at 30 C or 35 C, and is not complete after 16 days at 10 C; the oocysts are killed at 45 C. In sporulation, the sporont divides to form 2 sporoblasts, and each of these forms a sporocyst with 4 sporozoites and a residuum.

Merogony. Excystation occurs in 50–100 minutes at 22 C after treatment with trypsin and sodium taurocholate or in 1–2 hours after treatment with sodium taurocholate alone, but not after treatment with trypsin alone (Speer et al., 1973). At excystation, the sporocyst wall breaks, apparently at the places where the plates come together; in this it is quite different from *Eimeria* (Speer et al., 1973).

Merogony and gametogony were described by Lepp and Todd (1974). There are 3 generations of meront. The first-generation meronts mature in 5–7 days, are ellipsoidal to spherical, 16–38 x 11–23 (mean, 25 x 21) μm, and contain 4–24 (mean, 14) merozoites blunt at one end and rather pointed at the other, 8–11 x 3–5 (mean, 10 x 4) μm, with a central nucleus.

Second-generation meronts appear on day 6 and become mature in about 1 day. Mature meronts are spherical to slightly ovoid, 12–18 x 8–13 (mean, 15 x 11) μm, and contain 3–12 banana-shaped merozoites 11–12 x 3–5 (mean, 12 x 4) μm, with a central nucleus. The nuclei of some of them begin to divide before the merozoites leave the meront, developing to third-generation meronts there. Others leave the meront and enter new host cells before nuclear division.

The third-generation meronts become mature on days 7–8, mostly the latter. They are 13–38 x 8–24 (mean, 24 x 17) μm, and contain 6–72 (mean, 26) merozoites. These are not arranged in any particular order in the host cell. They are blunt at one end and rather pointed at the other, with a central nucleus and nucleolus; they are 8–13 x 1.5–3 (mean, 11 x 2) μm.

Gametogony. The third-generation merozoites enter new host cells and become spherical macrogametes and microgamonts. The earliest ones can be seen on day 7, but more are present on days 8–10. The mature macrogametes are ovoid, without distinct eosinophilic plastic granules, 22–29 x 14–23 (mean, 25 x 18) μm.

The microgamonts are 20–38 x 14–26 (mean, 29 x 20) μm, with a central residuum. They produce numerous microgametes about 5 x 0.8 μm.

A few oocysts can be seen on day 8 in tissue sections. They are 28–38 x 20–30 μm.

Prepatent Period. 9–11 (mean, 10) days (Neméseri, 1960; Lepp and Todd, 1974). Dubey (1975) found that the prepatent period in dogs fed oocysts was 9–11 days, while that in dogs fed infected mice was 8–9 days.

Patent Period. About 4 weeks (Neméseri, 1960).

Pathogenicity. There appear to be strain differences in pathogenicity. Neméseri (1960) found that about 5,000 oocysts of a Hungarian strain caused no signs of disease but that an inoculum of 50,000–80,000 oocysts caused dogs to be weak and to lose their appetites after 8–9 days. They then had diarrhea and dysentery and a temperature of 39.5–40.3 C. Their feces contained both blood and mucus. After about a week the signs subsided, and the animal gradually improved. Lepp and Todd (1974) apparently observed no signs in dogs fed as many as 200,000 oocysts of an Illinois strain.

Immunity. According to Lepp and Todd (1974), dogs given about 100,000 oocysts at 6 weeks of age were completely immune 2 months later, and dogs inoculated with about 200,000 oocysts at 10 weeks of age were completely immune 1 month after the end of patency. Hence *I. canis* apparently imparts a strong and lasting immunity.

Schrecke and Dürr (1970) found that pups 1 day old could be infected if enough oocysts were used. Susceptibility to infection increased slowly with age until weaning, when there was a marked increase. They thought that this was due mainly to diet-induced physiologic and biochemic properties of the intestine.

Cross-Transmission Studies. Loveless and Andersen (1975) transmitted *I. canis* from the dog to the coyote *C. latrans.* Kittens and either normal cats or those immunosuppressed with 6-methyl prednisolene acetate (Depomedrol, Upjohn) do not produce oocysts after having been fed *I. canis* oocysts (Neméseri, 1960; da Rocha and Lopes, 1971; Dubey, 1975). However, newborn kittens have the coccidia in their extraintestinal organs following oocyst ingestion, and puppies can become infected by eating these organs. Infected mice, too, do not shed oocysts, but can infect dogs if they are eaten (Dubey, 1975).

Levine and Cechner (1973) were unable to find *Sarcocystis* or parenteral stages of coccidia in baby pigs fed oocysts of *I. canis* from the dog.

Cultivation. Fayer and Mahrt (1972) cultivated *I. canis,* beginning with sporozoites, in embryonic bovine kidney, embryonic bovine trachea, and Madin-Darby kidney cell cultures and primary cultures of embryonic canine kidney and embryonic canine intestine. The sporozoites entered all types of cells and were abundant in all of them after 5 days except in the Madin-Darby canine kidney cells. None was found after 10 days in embryonic bovine kidney. The cells contained 1–14 sporozoites, which averaged 15 x 5 μm. They reproduced, apparently by endodyogeny (at least the daughter cells were in pairs), and there were occasionally clusters of merozoites in the rosettes. However, no further multiplication or reproduction took place. Mahrt (1973) saw both endodyogeny and schizogony in Madin-Darby bovine kidney cell cultures.

Jensen and Edgar (1978) studied the fine structure of the sporozoites as they penetrated the host cells in embryonic bovine trachea monolayer cultures. The host cell membrane remained intact during penetration, forming a vacuole within which the sporozoite lay. The rhoptries and micronemes, which were branched elements of the same network, often became empty or partially so during penetration.

Isospora ohioensis Dubey, 1975

(Plate 3, Figs. 13, 15, 18; Plate 4, Fig. 19)

Synonyms. Diplospora bigemina Wasielewski, 1904 from the dog in part; *Isospora rivolta* (Grassi, 1879) Wenyon, 1923 of *auctores* from the domestic dog; *Lucetina rivolta* (Grassi, 1879) Henry and Leblois, 1926 from the domestic dog; *Cystoisospora ohioensis* (Dubey, 1975) Frenkel, 1977; *Levinea ohioensis* (Dubey, 1975) Dubey, 1977.

Type Definitive Host. Domestic dog *Canis familiaris.*

Other Definitive Hosts. Coyote *C. latrans,* dingo *C. dingo* (?), fox *Vulpes vulpes* (?), perhaps raccoon dog *Nyctereutes procyonoides ussuriensis.*

Transport Host. House mouse *Mus musculus.*

Location. Epithelial cells throughout small intestine.

Geographic Distribution. Worldwide.

Prevalence. This species (or a combination of it, *I. burrowsi,* and *I. neorivolta*) has been found in 4%–72% of dogs (mean, 21%) in 7 surveys in the United States and Canada, and in 2%–12% in 2 surveys in other countries (Levine, 1973). It was found in 4% of 113 dogs in Ohio by Catcott (1946), 4% of 500 stray dogs in Ohio by Streitel and Dubey (1976), 17% of 139 dogs in Illinois by Levine and Ivens (1965a), 5% of 308 dogs in Illinois by Lepp and Todd (1974), 11% of 835 dogs in New Jersey by Burrows (1968), 22% of 155 dogs in Montreal by Choquette and Gelinas (1950), 13% of 45 dogs in Utrecht, Netherlands by Nieschulz (1925), 6% of 18 dogs in Brazil by Costa and Freitas (1959), 1% of 78 dogs in Chile by Barriga and Jaramillo (1966), 4% of 1505 dogs in Chile by Alcaino and Tagle (1970), 13% of 54 dogs in Iraq by Mirza (1970), 4% of 115 dogs in Iran by Mirzayans et al. (1972) and 2% of 39 dogs in Turkey by Mimioglu, Güralp, and Sayin (1960). It was reported by Arther and Post (1977) from 2% of 82 coyotes *C. latrans* in Colorado. Dubey, Fayer, and Seesee (1978) found *I. ohioensis*-like oocysts in the feces of 3% of 169 adult coyotes in Montana.

Oocyst Structure. Sporulated oocysts were described by Levine and Ivens (1965a) and Dubey (1975a). They are ellipsoidal to ovoid, 20–27 x 15–24 (mean, 23–24 x 19–21) μm, with a smooth, colorless to pale yellow wall composed of a single layer about 0.8 μm thick lined by a thin membrane, without micropyle, residuum, or polar granule. Sporocysts ellipsoidal, occasionally flattened on one side, 12–19 x 9–13 (mean, 15–17 x 10–12) μm, with wall about 0.4 μm thick, without Stieda body, with residuum. Sporozoites banana-shaped, about 9–13 x 2.5–5 (mean, 10 x 3.5) μm with 1 or more clear globules. Occasionally an abnormal oocyst containing a single sporocyst with 8 sporozoites or with 2–3 sporocysts with dissimilar numbers of sporozoites may be found. Upon refrigeration for several months, a high percentage of the oocysts seen by Levine and Ivens (1965a) had collapsed walls so that their sporocysts made them look like dumbbell-shaped doublets. At this time the sporocysts were ellipsoidal, with a pale yellowish wall about 0.6 μm thick and lay side by side. They were about 17 x 10 μm, without a Stieda body, with a residuum. The sporozoites were sausage-shaped, lay more or less lengthwise in the sporocysts, and contained 1 or more clear globules.

The oocysts reported by Arther and Post (1977) from coyotes were 17–23 x 15–21 (mean, 21 x 17) μm.

Merogony. Dubey (1978a) described the life cycle of *I. ohioensis* in dogs, using oocysts obtained from an infection with a single oocyst. All stages are in the epithelial cells of the intestine. The first-generation meronts were in the jejunum. Two days after inoculation, zoites 7–9 x 2.5 μm (in sections) were in pairs in parasitophorous vacuoles 7–9 x 6 μm. At 3 days multinucleated zoites and meronts containing fully developed merozoites were above the host cell nucleus in the villi of the jejunum. The meronts were 9–19 x 2.5–4 μm (in sections) and contained 2–8 merozoites 9–12 x 2.5–4 μm (in sections). Both uninucleate zoites, multinucleate elongate meronts, and fully formed merozoites were sometimes within the same parasitophorous vacuole.

Four to 5 days after inoculation, asexual multiplication was occurring throughout the small and large intestines, mostly in the ileum. Dubey (1978a) did not determine the number of asexual generations. He saw at least 2 structurally distinct meronts at this time. Type I meronts contained merozoites 11 x 3 μm, and Type II meronts contained merozoites 7.5 x 1.5 μm. He found uni-, bi-, and multinucleate zoites in the same parasitophorous vacuole.

Gametogony. Dubey (1978a) found gamonts in the surface epithelial cells of the small intestine, cecum, and colon, mostly the ileum, 4–5 days after inoculation. The macrogametes were 13–17 x 11–12 (mean, 14.5 x 13) μm in sections and 21–26 x 17–25 (mean, 22 x 18) μm in smears. The microgamonts were 13–17 x 8–15 (mean, 15 x 11) μm in sections and 24–30 x 15–24 (mean, 27 x 19) μm in smears, and contained up to 50 microgametes.

Prepatent Period. About 4.5 days after ingestion of oocysts or 3.5 days after ingestion of mouse transport hosts (Dubey, 1978a).

Patent Period. 3–5 weeks in pups infected at 6–10 days of age; 1–2 weeks in pups infected at 4–384 days of age (Dubey, 1978).

Pathogenicity. This species is usually nonpathogenic but is sometimes mildly pathogenic in dogs. Dubey (1978) found that 5 of 18 newborn pups fed either sporocysts or mouse tissues containing hypnozoites developed diarrhea 3–4 days after feeding. All of 13 weaned pups similarly inoculated remained healthy.

The coccidia were in the epithelial cells of the villi of the small intestine, especially the ileum. Histologic lesions observed were necrosis and desquamation of the tips of the villi of the ileum and atrophy of the villi (Dubey, 1978).

Immunity. Dubey (1978) found that pups more than 40 days old became immune within 1 week after oral inoculation and did not reshed oocysts after reinoculation. The immunity lasted 1–2 months.

Cross-Transmission Studies. Loveless and Andersen (1975) transmitted this species from the dog to the coyote *C. latrans.* Cats or mice infected with this species from the dog do not produce oocysts (Dubey, 1975a). However, Heydorn (1973) found that muscle, liver, lung, and brain tissue of mice that had been inoculated orally with *I. ohioensis* caused 4 germ-free dogs to shed oocysts in their feces from day 5 to days 27-33. Dubey (1975a) found that if kittens or mice were fed oocysts and then if the mice or the extraintestinal organs of these animals were fed to dogs, the dogs subsequently shed oocysts of *I. ohioensis*. Dubey and Mehlhorn (1978) found that the extraintestinal tissues of mice fed oocysts of *I. ohioensis* from the dog were infective for dogs beginning within 1 day of feeding and continuing for at least 211 days. The sporozoites grew in the mice from 5-6 μm to 11-16 μm in length by day 39 (i.e., they became hypnozoites).

Dubey (1978a) infected dogs by feeding them the mesenteric lymph nodes and spleens of mice that had been inoculated 17 days before, and he found hypnozoites in these organs. He also infected dogs by feeding them mesenteric lymph nodes and spleens of dogs that had been fed oocysts 5 days earlier, but he did not find any hypnozoites in these organs. To summarize, cats, dogs and mice are transport hosts for dogs.

Remarks. Thornton, Bell, and Reardon (1974) found what they called *I. rivolta* (but did not describe) in the feces of 1 of 2 adult coyotes *C. latrans* in Texas. Bearup (1954) found oocysts, which he called those of *I. rivolta*, in a captive dingo *(C. dingo)* pup in Australia. Gousseff (1933) found what he called most probably *I. rivolta* in the feces of foxes (presumably *V. vulpes* in Transcaucasia); the oocysts were spherical, 27-32 μm in diameter, without micropyle, residuum, or polar granule, with a smooth, colorless to bright orange wall 1-1.5 μm thick, with a sporocyst residuum; he did not give the dimensions of the sporocysts. Yakimoff and Matikaschwili (1933) gave the following information on a coccidium that they identified as *I. rivolta* from the raccoon dog *N. procyonoides ussuriensis* from fur farms in Voronezh and Novgorod, USSR. They were present in 7% of 241 fecal samples from Voronezh and 60% of 135 from Novgorod: oocysts ovoid, ellipsoidal, subspherical or spherical, the ovoid and ellipsoidal ones being 20-31 x 18-25 μm, the subspherical ones 19-28 x 18-25 μm, and the spherical ones 18-25 μm in diameter with a mean of 22 μm, without micropyle, residuum, or polar granule; sporocysts generally ellipsoidal, 14-18 x 9-13 (mean, 16 x 10.5) μm, without Stieda body, with residuum; sporozoites elongate with one end wider than the other, mean 11 x 4 μm; there were some abnormal oocysts containing a single sporocyst or 8 naked sporozoites, i.e., *Caryospora-* or *Tyzzeria*-like. Perhaps these were *I. ohioensis*. Because the original *I. ohioensis* has now been separated into 3 species (*I. ohioensis, I. burrowsi,* and *I. neorivolta*), the information given above on prevalence, hosts, and cross-transmission studies is somewhat uncertain.

Isospora neorivolta Dubey and Mahrt, 1978

(Plate 3, Fig. 14)

Synonym. Isospora rivolta (Grassi, 1879) Wenyon, 1923 of Mahrt (1967).
Type Definitive Host. Dog *Canis familiaris.*
Location. Posterior half of small intestine, rarely cecum and colon, in distal 1/3 of villi, mostly in the subepithelial cells of the lamina propria but occasionally in the epithelial cells (Mahrt, 1967).
Geographic Distribution. North America (Illinois).
Prevalence. Unknown; see *I. ohioensis.*
Oocyst Structure. Structural details and measurements of the oocysts of this species have apparently not been published. However, they are similar to those of *I. ohioensis* (Dubey and Mahrt, 1978).
Sporulation. The oocysts are unsporulated when passed in the feces. Complete sporulation takes 48 hours at 20 C, 24 hours at 25 C, 16 hours at 30 C, or 8 hours at 38 C; the oocysts do not develop or survive at 50 C (Mahrt 1968).
Merogony. Mahrt (1967) described the endogenous stages after infecting puppies with a cloned culture derived from a single oocyst. There are at least 2 generations of meronts and merozoites. The meronts are 17–25 x 12–25 µm and contain 4–24 merozoites (most commonly, 4 or 8) each. He found meronts containing mature merozoites as early as 3 days after inoculation of oocysts; they reached a maximum 4 days after inoculation. They were slender, curved, and 10–13 x 2–3 µm.

Dubey and Mahrt (1978) found at least 4 types of meront: type I meronts occurred 72 hours after inoculation. They were 6–10 x 2–3 (mean, 8 x 2.5) µm, and contained 1–8 merozoites. Type II meronts occurred in the ileum 96-120 hours after inoculation. They were 9-18 x 7-18 μm and contained 2-12 short, broad merozoites 8-12 x 3-5 (mean, 9 x 3.5) μm. Most of these merozoites were binucleate, but others had 1 or 3 nuclei; mono- and binucleate merozoites occurred together within the same host cell vacuole. Type III meronts were 7-25 x 4-16 (mean, 13 x 8) μm and contained 2-20 mononucleate merozoites 8-14 x 2-3 (mean, 10 x 2.5) μm, and occasionally an eosinophilic residuum if the merozoites formed a rosette. This type of meront was seen predominantly at 96 hours. Type IV meronts were 9-18 x 7-18 μm and contained 4-30 short, thin merozoites 7-8 x 1-1.5 (mean, 7.5 x 1) μm. Type IV meronts were seen mainly at 120-144 hours.

Gametogony. According to Mahrt (1967), mature macrogametes and microgamonts were present 6 days after oral inoculation of oocysts. The mature macrogametes contained eosinophilic plastic granules. The mature microgamonts averaged 13 x 9 μm and contained 50-70 microgametes with slightly curved, tapering bodies 6 x 0.6 μm and with 2 posteriorly directed flagella 11-14 μm long.

Prepatent Period. 142-146 hours after feeding oocysts (Mahrt, 1967).

Patent Period. 13-23 (mean, 19) days (Mahrt, 1967).

Pathogenicity. This species is apparently only slightly pathogenic (Mahrt, 1966).

Remarks. This species was separated from *I. ohioensis* by Dubey and Mahrt (1978) because its endogenous stages develop predominantly in the lamina propria of the posterior half of the small intestine and rarely in the cecum and colon, whereas the endogenous stages of *I. ohioensis* occur only in the epithelial cells of the villi, and throughout the small intestine. In addition, multinucleated merozoites containing 7-8 nuclei are numerous in *I. ohioensis,* whereas they are absent in *I. neorivolta.*

Isospora arctopitheci Rodhain, 1933

Hendricks (1977) said that he had transmitted this monkey species from experimentally infected marmosets *Saguinus geoffroyi* to the dog *Canis familiaris,* coatimundi *Nasua nasua,* kinkajou *Potos flavus,* tayra *Eira barbara,* and cat *Felis catus,* and that he had transmitted it back to *S. geoffroyi* from most of them.

Isospora sp. Dubey, Weisbrode, and Rogers, 1978

Type Host. Dog *Canis familiaris.*
Location. Villar epithelium, lamina propria, and intestinal glands of the distal half of the ileum, cecum, and colon.
Geographic Distribution. North America (Ohio).
Oocyst Structure. Similar to that of *I. ohioensis;* not measured in feces or described because culture was inadvertently discarded. Oocysts 16–23 x 14–20 (mean, 19 x 16) μm in smears and 12–17 x 10–13 (mean, 13 x 11.5) μm in sections.
Sporulation. Occurs outside host's body.
Merogony. Dubey, Weisbrode, and Rogers (1978) said that there were 2 or more generations of meront. The meronts were generally merozoite- (spindle-) shaped and contained up to 7 nuclei. They saw at least 3 sizes of uninucleated free merozoites in smears—A merozoites 8–10 x 1.5–2 (mean, 9 x 2) μm; B merozoites 12–15 x 1.5–2 (mean, 13 x 2) μm; and C merozoites 12–16 x 4–6 (mean, 14 x 4.5) μm. Binucleated meronts were 15–16 x 2–3 (mean, 16 x 2) μm, 3 nucleated meronts were 15–18 x 3–4 (mean, 16 x 3) μm, and 4–6 nucleated meronts were 18–19 x 4–6 (mean, 18.5 x 5) μm. Meronts were more numerous than gamonts. "Regular" meronts were oval to elliptical and contained up to 30 merozoites. There were at least 2 structurally different meronts, which were distinguished by having different sizes of merozoites. The merozoites contained a conoid, rhoptries, Golgi apparatus, microtubules, endoplasmic reticulum, and mitochondria.
Gametogony. Early gamonts were merozoite-shaped. Microgamonts contained up to 45 nuclei. There were at least 5 times as many macrogametes as microgamonts.
Pathogenicity. Dubey, Weisbrode, and Rogers (1978) found this coccidium in a fatal case of coccidiosis in a 10-week-old pup. This seems to be the only species of coccidium especially pathogenic for dogs.
Remarks. Dubey, Weisbrode, and Rogers (1978) said that this might be a mixture of *I. ohioensis* and *I. neorivolta,* but concluded

that it was an unidentified coccidium with oocysts structurally similar to those of *I. ohioensis*. It differed from *I. ohioensis* in pathogenesis and also in that *I. ohioensis* occurs only in the villar epithelium and not in the lamina propria or intestinal glands. They could not study it further because it had been inadvertently discarded.

Sarcocystis cruzi (Hasselmann, 1926) Wenyon, 1926

(Plate 1, Fig. 6)

Synonyms. Miescheria cruzi Hasselmann, 1926 in part; *Sarcocystis fusiformis* Railliet, 1897 of Babudieri (1932) and *auctores*, in part; *[non] S. fusiformis* (Railliet, 1897) Bernard and Bauche, 1912; probably *Sarcocystis iturbei* Vogelsang, 1938; *Sarcocystis marcovi* Vershinin, 1975 in part; *Sarcocystis bovicanis* Heydorn et al., 1975; "free sporocysts of *Isospora rivolta*" of Gassner (1940), Levine and Ivens (1965a) and *auctores* in part; "large form of *Isospora bigemina*" of Mehlhorn, Heydorn and Gestrich (1975) and Heydorn, Mehlhorn and Gestrich (1975) in part; "?Cryptosporidium" of Bearup (1954) in part; probably *Sarcocystis* sp. Golemansky and Ridzhakov (1975) of red fox *Vulpes vulpes*; probably *Cryptosporidium vulpis* Wetzel, 1938; *Endorimospora hirsuta* (Moulé, 1888) Tadros and Laarman, 1976.

Type Definitive Host. Domestic dog *Canis familiaris*.

Other Definitive Hosts. Coyote *C. latrans*, wolf *C. lupus*, red fox *Vulpes vulpes*, raccoon *Procyon lotor*.

Intermediate Host. Ox *Bos taurus*.

Location. Gamonts, gametes, zygotes, oocysts, and sporocysts in lamina propria of villi of small intestine, especially distal jejunum and proximal ileum (Fayer, Mahrt, and Johnson, 1973; Fayer, 1974) of dogs. First-generation meronts in endothelial cells of arteries in cecum, large intestine, kidney, pancreas, and cerebrum (Fayer, 1977a). Second-generation meronts in endothelial cells of capillaries of many organs, mostly in kidney glomeruli (Fayer and Johnson, 1973, 1974). Sarcocysts in skeletal muscles (esophagus, diaphragm, heart, other muscles of ox (Fayer, Johnson, and Hildebrandt, 1976).

Geographic Distribution. Worldwide.

Prevalence. Common. According to Levine (1973), *Sarcocystis* (presumably both *S. cruzi, S. hirsuta,* and *S. hominis*) was found in 75%–98% (mean, 85%) of cattle in 3 surveys in the United States. Skibsted (1945) found it in 94% of 100 cows and 18.5% of 97 calves in Denmark. Farmer et al. (1978) found *Sarcocystis* sporocysts in 36.5% of 123 sheepdogs from Wales and identified *S. cruzi* in 13 out of 47 of the sheepdogs. Using tryptic digestion, Boch, Laupheimer, and Erber (1978) found that 99.7% of 1,020 cattle from 11 slaughter houses in south Germany were infected with *Sarcocystis* spp.; using the trichinoscope they found only 57.9%. They found *S. cruzi* in 65.6% of 817 positive cattle. Feeding of muscle from unselected cattle to dogs ordinarily resulted in infection of the dogs.

Oocyst Structure. The oocysts sporulate in the host. They are spherical before sporulation, but become dumbbell-shaped, with the thin wall stretched between the 2 sporocysts after sporulation. They lack a micropyle, residuum, and polar granule. The sporocysts are ellipsoidal, with one side flatter than the other, 13–22 x 6–15 (mean, 15–17 x 8–11) μm, with a smooth, colorless to very pale yellow wall about 0.5 μm thick, without a Stieda body, with a residuum. The sporozoites are banana-shaped, with one end rounded and the other bluntly pointed, about 11 x 2–3 μm, usually with a clear globule near the wide end, and lie lengthwise in the sporocysts (Levine and Ivens, 1965a; Heydorn and Rommel, 1972; Mahrt, 1973; Suteu and Coman, 1973; Fayer and Leek, 1973, Fayer, 1974; Rommel et al., 1974; Fayer and Johnson, 1975; Gestrich, Heydorn, and Baysu, 1975).

Sporulation. Occurs in the cells of the definitive host. Fayer (1974) found that it began in the dog 8 days after ingesting sarcocysts from cattle. The feces contain free sporulated sporocysts and a few sporulated oocysts. Leek and Fayer (1979) found that *S. cruzi* sporocysts survived without appreciable deaths for over 300 days under refrigeration in distilled water or in Hanks's balanced salt solution with penicillin, streptomycin, Fungizone, and Mycostatin, but were for the most part quickly killed (within a few days) under refrigeration in 2% sulfuric acid, 2.5% potassium bichromate, or 1% sodium hypochlorite solutions. They survived well at room temperature in distilled water in one experiment, and moderately well in another. They concluded that the sporocyst wall may be much more vulnerable to chemical disinfectants than that of *Eimeria* or *Isospora*.

Merogony. There are at least 3 generations of meront in cattle. The first-generation meronts were found by Fayer (1977b) in endothelial cells of small and medium-sized arteries in the cecum, large intestine, kidney, pancreas, and cerebrum of calves killed 15–16 days after oral inoculation with sporocysts from dogs. They were relatively scarce. They were 17–52 x 7–28 (mean, 29 x 11) μm and contained 8–200 (mean, 103) nuclei. None was mature, i.e., contained merozoites. The second-generation meronts were found by Fayer and Johnson (1973, 1974) in calves fed sporocysts 26–33 days before. They were in endothelial cells of the capillaries in many organs, especially the kidney glomeruli. They varied greatly in size; 25 of them were 8–27 x 4–13 (mean, 15 x 9) μm; they contained 3–50 nuclei (mean, 27). A few appeared nearly mature and contained nuclei associated with cytoplasm, but most were immature. According to Mehlhorn, Heydorn, and Gestrich (1975), calves had numerous meronts and merozoites 27 days after inoculation; these were in the liver, lung, kidney, heart, small intestine, esophagus, cerebrum, cerebellum, skeletal muscles, diaphragm, and other organs. The merozoites were 7–8 x 2–3 μm. Endodyogeny was also present.

The last-generation meronts (the sarcocysts) are large and occur in striated muscles. Beginning on day 34, Mehlhorn, Heydorn, and Gestrich (1975) found numerous "cysts" containing small numbers of metrocytes only. These "cysts" developed in a parasitophorous vacuole in the host cells. At first the vacuole had a single unit membrane, but it thickened to form a primary wall up to 20–25 nm thick. As the "cyst" grew, this primary wall became folded in alternating long and short club-shaped protrusions. The long ones were about 0.6 μm long; the short ones were about 0.13 μm long. The combined protrusions looked like a very thin cyst wall under the light microscope. Later all of the protrusions lengthened, becoming about 3 μm in maximum length. They did not contain fibrils. Even old cysts appeared to be relatively thin-walled under the light microscope because the protrusion layer was not thicker than 1 μm due to folding over of the protrusions. As the cyst grew, it became progressively more condensed due to the development of an amorphous ground substance containing fine fibrils and granules. It became divided by thin septa (not visible with the light microscope) into numerous chambers filled with parasites. Fayer and Johnson (1974) found sarcocysts containing merozoites or metrocytes (but no early meronts) in the diaphragm, esophagus, heart, skeletal muscles, and tongue 40–54 days after inoculation. At first there were only metrocytes in the sarcocysts according to Mehlhorn, Heydorn, and Gestrich (1975), but at 76 days and later only merozoites were present. Both metrocytes and merozoites reproduced by endodyogeny. They concluded that it took about 3 months for the sarcocysts to become fully sporulated and ready for transmission to dogs. Complete development of the sarcocysts took place in muscle cells. The parasitized cells were never surrounded by a fibrillar layer of host origin (i.e., had no secondary cyst wall). Mehlhorn et al. (1975) and Mehlhorn, Hartley, and Heydorn (1976) found that the sarcocysts had relatively few protrusions compared to *S. hirsuta* and *S. hominis* and that these were always folded over to lie on the surface and never contained microfibrils or microtubules.

The sarcocysts are up to 1 cm or more long. They are compartmented when mature and contain merozoites about 10 μm long.

Simpson (1966) described the fine structure of the sarcocysts of this species (presumably also of *S. hirsuta* and *S. hominis*, since his work was done before it was known that cattle had 3 species of *Sarcocystis*) in the myocardium of cattle. The sarcocyst wall appears to be entirely a product of the parasite. It is composed of a single layer about 200 nm thick, which extends into the sarcocyst and divides it into compartments. Each merozoite contains 3 fairly well-defined regions. The anterior 1/3 contains rather regular rows of micronemes about 50 nm in diameter, which seem to originate in the conoid. The middle 1/3 contains electron-dense granular bodies up to 0.4 μm in diameter, homogeneous bodies, often vacuolated, up to 0.3 μm in diameter, mitochondria, and many free ribosomes. The posterior 1/3 contains a nucleus with a prominent nucleolus and a double membrane containing nuclear pores, a few dense, oval osmiophilic bodies, many free ribosomes, and elongated mitochondria containing convoluted tubules. The merozoites have 2 membranes, each about 8 nm thick, about 20 nm apart.

Prepatent Period. 8–33 days in domestic dog (Heydorn and Rommel, 1972; Fayer and Leek, 1973; Suteu and Coman, 1973;

Fayer, Mahrt, and Johnson, 1973; Fayer, 1974, 1977; Gestrich, Heydorn, and Baysu, 1975; Dubey and Streitel, 1976); 8-9 days in dog, fox, and wolf (Rommel et al., 1974); or 9-10 days in coyote pups (Fayer and Johnson, 1975).

Patent Period. 3-70 or more days in dog (Heydorn and Rommel, 1972; Fayer, 1974, 1977; Fayer and Leek, 1973; Suteu and Coman, 1973; Rommel et al., 1974; Gestrich, Heydorn, and Baysu, 1975; Dubey and Streitel, 1976; Fayer, Johnson, and Hildebrandt, 1976); 14-21 days in fox (Fayer, Johnson, and Hildebrandt, 1976); 8-22 days in raccoon (Fayer, Johnson, and Hildebrandt, 1976); 10-11 days in coyote pups (Fayer and Johnson, 1975). Most sporocysts were shed by dogs from days 15-30 after being fed sarcocysts, and peak numbers were shed on days 23 and 24 (Fayer, 1977).

Pathogenicity. So far as is known, this species is not pathogenic for carnivores. However, it is highly pathogenic for the ox. Fayer and Johnson (1973, 1974) found that 4 of 5 calves fed 250,000-1,000,000 sporulated sporocysts from dog feces died or became recumbent 26-35 days later. The calves had anorexia, cachexia, weight loss, anemia, and accelerated heart rates 23-33 days after inoculation. They were killed or died 26-54 days after inoculation (Fayer, Mahrt, and Johnson, 1973) and were found to have hemorrhage of the serosa throughout the peritoneal cavity, hemorrhage of the pericardium, myocardium, dorsal surface of the cerebellum, and generalized lymphadenopathy. Typical coccidian meronts were found by histologic examination of various organs (listed under *Merogony* above) up to 33 days after inoculation; thereafter typical sarcocysts were found in the striated muscles but there were no meronts in the organs. Control calves remained negative. Mahrt and Fayer (1975) found that *S. cruzi* caused an oligocythemic anemia, leukocytic shift to the left, and elevated serum SGOT, LDH, and CPK levels during the acute phase of the disease in experimentally infected calves. Fayer, Johnson, and Lunde (1976) reported abortion and death in experimentally infected adult cattle. Frelier et al. (1977) observed illness and death of cattle under field conditions.

Gestrich, Heydorn, and Baysu (1975) found that 5 calves each fed 2 million *S. cruzi* sporocysts from dogs died 27-30 days later. There were a great many merozoites and meronts in their organs.

Low doses of sporocysts did not cause death, but produced thin-walled (less than 0.5 μm thick) sarcocysts, which were found in the muscles of the calves on examination 98 days later.

Markus, Killick-Kendrick, and Garnham (1974) considered that Dalmeny disease in Canadian cattle described by Corner et al. (1963) was due to *Sarcocystis* meronts. In this outbreak 25 cattle were affected; 5 of them died and 12 were killed when moribund; 10 of the 17 pregnant cows among them aborted. Small meronts indistinguishable in retrospect from those of *Sarcocystis* were found in the endothelial cells of the blood vessels of many organs in 11 out of 16 of the animals examined histologically. Lainson (1972) found a similar organism in the liver and lungs of a heifer calf in England, and Markus, Killick-Kendrick, and Garnham (1974) thought it was the same. It seems certain that Dalmeny disease and the English case were both due to *Sarcocystis*. Schmitz and Wolf (1977) reported a fatal case, possibly due to *S. cruzi*, in a 2-week-old calf in Oregon. Meronts 10–24 x 4–19 μm were present in the kidneys and also in the lungs, and a few were in the myocardium.

Fayer (1974) and Fayer and Johnson (1975) produced a similar disease in calves fed sporocysts of *S. cruzi* from naturally infected coyotes from Utah and Idaho. One calf fed 135,000 sporocysts became recumbent and was killed 43 days after inoculation. Another calf, fed 720,000 sporocysts, died 50 days after inoculation. Both calves had gross lesions of mild hydroperitoneum and hydropericardium, pale musculature, and pale mucous membranes; the second calf also had mild hydrothorax, petechiation, and mottling of the ventricular myocardium, dark red discolorations on the lungs, and enlarged lymph nodes. Both calves had mild mononuclear infiltration of the connective tissues and moderate lymphoid hyperplasia of the spleen and lymph nodes. There were young meronts in the lymph nodes or muscles.

Immunology. Aryeetey and Piekarski (1976) considered the indirect immunofluorescent test (IIFT) reliable in detecting *Sarcocystis* infection.

Lunde and Fayer (1977) found that the indirect hemagglutination (IHA) titer of experimentally exposed cattle (against a *S. cruzi* soluble antigen prepared from bradyzoites in sarcocysts in the heart muscle) began to rise 30–45 days after inoculation and became as high as 1:39,000 90 days after inoculation. Because dairy cows from the field had titers as high as 1:486, they concluded that such titers were probably not significant for diagnosis. They also found precipitins in the gel diffusion test beginning 30 days after inoculation. They found no cross-reaction in the IHA test between *S. cruzi* and *Toxoplasma gondii* with human sera.

Cross-Transmission Studies. Cross-transmission between predators by feeding sporocysts does not take place. It can occur only if the sporocysts are fed to the ox and the ox tissues containing sarcocysts are then fed to the predators. By this means it has been found that the wolf *C. lupus* (Rommel et al., 1974), red fox *V. vulpes* (Rommel et al., 1974; Fayer, Johnson, and Hildebrandt, 1976), coyote *C. latrans* (Fayer, 1974; Fayer and Johnson, 1975) and raccoon *P. lotor* (Fayer, Johnson, and Hildebrandt, 1976) can be infected. The first 3 belong to the same family as the dog, but the raccoon belongs to a different one.

Fayer, Johnson, and Hildebrandt (1976) infected the dog via the ox with *S. cruzi* from the coyote.

Animals that cannot be infected (or at least that do not pass oocysts or sporocysts after feeding bovine tissues infected with sarcocysts) are the cat (Gestrich, Heydorn, and Baysu, 1975; Fayer, Johnson, and Hildebrandt, 1976), skunk *Mephitis mephitis* (Fayer, Johnson, and Hildebrandt, 1976), ferret *Mustela furo* (Fayer, Johnson, and Hildebrandt, 1976), hyena *Crocuta crocuta* (Rommel et al., 1974), brown bear *Ursus arctos* (Rommel et al., 1974), pig (Fayer, Johnson, and Hildebrandt, 1976), sheep (Gestrich, Schmitt, and Heydorn, 1974; Gestrich, Heydorn, and Baysu, 1975; Fayer, Johnson, and Hildebrandt, 1976), rhesus monkey (Fayer, Johnson, and Hildebrandt, 1976), laboratory rabbit (Fayer, Johnson, and Hildebrandt, 1976), guinea pig (Suteu and Coman, 1973; Fayer, Johnson, and Hildebrandt, 1976), laboratory mouse (Suteu and Coman, 1973), laboratory rat (Aryeetey and Piekarski, 1976; Fayer, Johnson, and Hildebrandt, 1976), and chicken (Suteu and Coman, 1973).

In addition to not passing sporocysts, Aryeetey and Piekarski (1976) found that the laboratory rat did not become positive to the IIFT after eating infected meat from cattle or different kinds of sausages.

Remarks. Tadros and Laarman (1976) used the name *S. hirsuta* for the ox-dog species *S. cruzi.* They said that if one dissects away the capsule of the sarcocyst with fine needles, the outer wall has a hirsute appearance. However, the ox-cat form (which we consider to be *S. hirsuta*) has this appearance without removal of the capsule, while the ox-dog form does not (see Rommel, 1975), and there is no evidence that Moulé (1888) removed the capsule when he named *S. hirsuta*.

So far as is known, the sporocysts of this species cannot be distinguished from those of *S. tenella, S. miescheriana,* or *S. bertrami,* which also occur in the dog.

Perhaps this is the species that Bigalke and Tustin (1960) found in the cerebellum of an ox in South Africa, that Luengo, Arota, and Luengo (1974) found in the cerebellum of a heifer that died of tubercular meningoencephalitis in Chile and that Clegg, Beverley, and Markson (1978) found causing unthriftiness in calves in England.

Yakimoff and Matchulski (1935) found sporocysts that they called *I. rivolta* in the dingo *C. dingo* in the Leningrad zoo. They said that they were 13–18 x 9–13 (mean, 16 x 11) μm, with a transparent, double-contoured wall, and a residuum. These sporocysts were undoubtedly those of *Sarcocystis.* Bearup (1954) found cysts in the feces of a captive dingo pup in Sydney, Australia, that he thought were probably sporocysts of *I. rivolta* or *Cryptosporidium* sp. They averaged 17 x 11 μm and contained 4 sausage-shaped sporozoites averaging 11 x 2 μm and a residuum.

Yakimoff and Matchulski (1935) found a form that they also called *I. rivolta* in the feces of a wolf *C. lupus* from northern USSR in the Leningrad zoo. They said that the oocysts were irregularly ovoid, 14–18 x 11–13 (mean, 15.5 x 11) μm and that they were already sporulated in fresh feces. They did not give sporocyst dimensions but said merely that the sporozoites were 9–10 x 2–3 μm and that a sporocyst residuum was present. What they spoke of as oocysts may have been actually free sporocysts, and they may actually have been dealing with *Sarcocystis* in this animal also.

Mackinnon and Dibb (1938) found sporocysts that they said agreed very well with those of *I. bigemina* of the dog in the feces of an Arctic fox (which they called *C. occidentalis* but which we suspect was *Alopex lagopus*) in the London zoo. The sporocysts contained 4 sporozoites, which were 14–15.5 x 9–10 μm. This was probably *Sarcocystis.*

Mackinnon and Dibb (1938) found sporocysts of an *Isospora* that they considered to be *I. rivolta* in the feces of a raccoon dog *Nyctereutes procyonoides* in the London zoo. They were 14–18 x 10–12 μm and contained 4 sporozoites. They were undoubtedly *Sarcocystis*.

Sarcocystis tenella (Railliet, 1886) Moulé, 1886

(Plate 1, Fig. 2)

Synonyms. Miescheria tenella Railliet, 1886; *Sarcocystis ovicanis* Heydorn et al., 1975; "free sporocysts of *Isospora rivolta*" of Gassner (1940), Levine and Ivens (1965a) and *auctores* in part; "large form of *Isospora bigemina*" of Mehlhorn, Heydorn, and Gestrich (1975) and Heydorn, Mehlhorn, and Gestrich (1975) in part; probably "?Cryptosporidium" of Bearup (1954) in part; probably *Hoareosporidium pellerdyi* Pande, Bhatia, and Chauhan, 1972; *Endorimospora ovicanis* (Heydorn et al., 1975) Tadros and Laarman, 1976. (Ashford, 1977 considered *Isospora bigemina* to be a synonym of *S. tenella*.)

Type Definitive Host. Domestic dog *Canis familiaris*.

Other Definitive Host. Red fox *Vulpes vulpes*, coyote *C. latrans*, probably *C. dingo*.

Intermediate Host. Domestic sheep *Ovis aries*.

Location. Gamonts, gametes, zygotes, oocysts, and sporocysts most numerous in subepithelial tissue at tips of villi in proximal 1/3 of small intestine of dog (Munday, Barker, and Rickard, 1975). Earlier meronts in endothelium of arteries, arterioles of many organs, including brain of sheep. Late meronts (sarcocysts) in striated muscles of sheep.

Geographic Distribution. Worldwide.

Prevalence. Common. Ashford (1977) found sporocysts of this species in 12 out of 22 wild red foxes *V. vulpes* in England. Farmer et al. (1978) found *Sarcocystis* spp. sporocysts in 36.5% of 123 sheepdogs from Wales, 24% of 33 racing greyhounds from London, and 17% of 41 foxes from Wales. On the basis of sporocyst size, they identified *S. tenella* in 30 of 47 sheepdogs, 6 of 8 greyhounds, and 5 of 5 foxes. Dubey, Fayer, and Seesee (1978) found *"Sarcoc-ystis* sp." sporocysts in 53% of 169 adult coyotes in Montana.

Oocyst Structure. The oocysts are spherical before sporulation, but become dumbbell-shaped, with a thin wall stretched between the 2 sporocysts after sporulation. They lack a micropyle, residuum, and polar granule. The sporocysts are ellipsoidal, with one side flatter than the other, 13–16 x 8–11 (mean, 14–15 x 9–10) μm, with a smooth, colorless to very pale yellow wall about 0.5-μm thick, without a Stieda body, with a residuum. The sporozoites are banana-shaped, with one end rounded and the other bluntly pointed, about 11 x 2–3 μm, lying lengthwise in the sporocysts, and usually with a clear globule near the wide end (Levine and Ivens, 1965a; Rommel et al., 1974; Munday and Rickard, 1974; Munday, Barker, and Rickard, 1975; Heydorn et al., 1975). The sporocysts reported by Ashford (1977) from the red fox were 13–14 x 9–10 μm.

Sporulation. Occurs in the intestinal cells of the definitive host. Munday, Barker, and Rickard (1975) saw it beginning 7–10 days after dogs had been fed sarcocysts from sheep. The first nuclear division produces 2 polar nuclei which divide laterally, producing 2 sporocysts each with 2 polar nuclei. Nuclear division is repeated to produce 4 sporozoites in each sporocyst. They found many pairs of sporulated sporocysts beneath the epithelium at the tips of the villi of dogs killed during patency. The feces contain mostly free sporulated sporocysts and a few sporulated oocysts.

Ashford (1977) found sporulated oocysts in the subepithelial tissue of the small intestine villi throughout its length, but mostly in the anterior part. The sporocysts were 13 x 9 μm on release, elongate ovoid with one side slightly flattened, with a residuum but without a Stieda body.

Merogony. There are probably 3 generations of meronts in sheep. The first generation is small. Munday, Barker, and Rickard (1975) found meronts in the endothelium of arteries and arterioles in many organs (but not the brain) of a lamb 15 days after feeding it sporocysts from the dog. They found smaller meronts in the capillary endothelium of many organs, including the brain, of another lamb 24 days after it had been fed sporocysts from a dog. Gestrich, Schmitt, and Heydorn (1974) found meronts and merozoites in touch smears of the liver, kidney, spleen, heart, lung, lymph nodes, muscles, diaphragm, esophagus, tongue muscles, and small intestine wall in 2 lambs that had been fed 1.6–2.0 million sporocysts from dog feces and that had died 24–25 days later. Heydorn and Gestrich (1976) found merozoites in the blood and meronts in the kidneys of a lamb that had died 25 days after having been fed 2 million sporocysts from a dog. They found only immature sarcocysts containing metrocytes in the muscles of another lamb that had remained apparently healthy after having been fed 100,00 sporocysts and that had been killed 41 days later. They found mature compartmented sarcocysts containing both metrocytes peripherally and infectious merozoites more centrally. Mehlhorn, Heydorn, and Gestrich (1975) studied the structure of the muscle meronts (sarcocysts) in the muscles of sheep killed 41, 63, and 81 days after having been fed sporocysts from the dog. The meronts were always in muscle fibers that were never surrounded by fibrillar layers (i.e., there was no secondary, host-generated meront wall). The meront was limited by a unit membrane that was thickened at many places of its interior by osmiophilic material. They called this complex, which was up to 25 nm thick, the primary cyst wall. In old meronts it was folded regularly to form palisade-like protrusions about 3.5 μm long, which contained neither microfibrils nor microtubules. With light microscopy the combined protrusions looked like a radially striated "thick wall." Mehlhorn, Hartley, and Heydorn (1976) extended these observations. Beneath the primary cyst wall was a zone of fine granular ground substance that extended as septa to the interior of the sarcocyst, dividing it into compartments.

The sarcocysts of *S. tenella* are always microscopic, a characteristic that differentiates them from the macroscopic sarcocysts of *S. gigantea,* which also occur in sheep muscles. Railliet (1886) said that the sarcocysts were 500 μm x 60–100 μm.

Gametogony. There is no merogony in the definitive host. Munday, Barker, and Rickard (1975) followed the process of gametogony in dogs killed periodically after having been fed mutton containing microscopic sarcocysts. They found macrogametes and microgamonts with peripherally developing microgametes in the proximal third of the small intestine 1 day after inoculation. Becker,

Mehlhorn, and Heydorn (1979) said that sexual stages are formed rapidly, the process being done in a day, at which time they saw oocysts.

Prepatent Period. In the dog, 8–14 days (Rommel et al., 1974; Heydorn et al., 1975; Munday, Barker, and Rickard, 1975; Dubey and Streitel, 1976) or 10–36 days (Leek, Fayer, and Johnson, 1977); in the red fox, 9–10 days (Ashford, 1977).

Patent Period. More than 9 days in the dog (Dubey and Streitel, 1976); more than 30 days in the red fox (Ashford, 1977).

Pathogenicity. This species is highly pathogenic for lambs. Gestrich, Schmitt, and Heydorn (1974) found that 2 8-week-old lambs fed 1.6–2.0 million sporocysts of *S. tenella* from dogs had fever, anemia, and inappetance 14 days later and died 24–25 days after feeding. Sections of their organs and tissues looked like those of calves that had died after infection with *S. cruzi* sporocysts from the dog. Gestrich, Heydorn, and Baysu (1975) reported the same thing of 3 lambs fed 2 million *S. tenella* sporocysts each, as did Heydorn and Gestrich (1976) of 1 lamb. The last reported that a lamb given 100,000 sporocysts died on day 29. Munday, Barker, and Rickard (1975) reported that 2 lambs fed sporocysts died 42 and 104 days later after an illness characterized by anemia and poor condition. They found mature meronts in cells of the brain 42 days after inoculation; they were associated with nonsuppurative meningoencephalitis. They also found developing sarcocysts associated with myositis in the muscles of this lamb. They found mature sarcocysts in the muscles of the lamb that died 104 days after after inoculation, and degenerate and mature sarcocysts together with nonsuppurative meningoencephalitis in the brain.

Leek, Fayer, and Johnson (1977) observed anemia, inappetence, weight loss, fever, and reduced serum protein in lambs infected by feeding sporocysts from dogs. The lambs that had been given 100,000 sporocysts (the smallest number given) died 27–29 days after inoculation, and those that had been given 1 million oocysts died 24–25 days after inoculation. The most apparent gross lesion was hemorrhage of the striated muscles and visceral organs, the heart being the most severely affected.

Pande, Bhatia, and Chauhan (1972) said that *"Hoareospiridium pellerdyi"* was mildly pathogenic in the dog, causing aggregations of erythrocytes inside the villar cores and suggestions of congestion in the mucosa of the attacked region.

S. tenella is apparently not pathogenic in the fox.

Cross-Transmission Studies. S. tenella is not infectious for the cat (Rommel, Heydorn, and Gruber, 1972) or the ox (Gestrich, Heydorn, and Baysu, 1975). Dubey, Fayer, and Seesee (1978) transmitted *"Sarcocystis* sp." (probably *S. tenella*) from the coyote to the sheep and then to the dog. It caused anemia in the sheep.

Cultivation. Mehlhorn, Becker, and Heydorn (1978) cultivated this species from sheep sarcocysts in dog kidney but not human fibroblast or cat lung cells. Becker, Mehlhorn, and Heydorn (1979) added pig kidney cells to the negative list and said that both oocysts and sporocysts developed in dog kidney cells.

Remarks. In their initial study of sheep *Sarcocystis,* Rommel, Heydorn, and Gruber (1972) found only *S. gigantea* (whose definitive host is the cat) in sheep, but Ford (1974, 1975) believed that, because of their close association, the dog must be a more common definitive host than the cat in Australia, and he discovered *S. tenella* there. It has subsequently been found in Europe and North America.

So far as is known, the sporocysts of this species apparently cannot be distinguished from those of *S. cruzi, S. miescheriana,* or *S. bertrami,* which also occur in the dog. Ashford (1977) said that they could not be distinguished from those of *I. bigemina* (which he considered a synonym of *S. tenella,* but which we consider more probably a synonym of *S. miescheriana*).

Perhaps this is the species that Hartley and Blakemore (1974) found associated with severe encephalomyelitis and myelomalacia in 2 young sheep in Australia. The merozoites in its sarcocysts were 5-7 μm long. This is also perhaps the species that Hilgenfeld and Punke (1974) found in the brain of a sheep in Germany.

Sarcocystis miescheriana (Kühn, 1865) Labbé, 1899

Synonyms. Synchrytium miescherianum Kühn, 1865; *Sarcocystis miescheri* Lankester, 1882; *Sarcocystis suicanis* Erber, 1977; *Coccidium bigeminum* Stiles, 1891; *Coccidium bigemina* var. *canis* Railliet and Lucet, 1891; *Isospora rivolta* (Stiles, 1891) Lühe, 1906; *"Isospora rivolta"* sporocysts of Gassner (1940), Levine and Ivens (1965a) and *auctores* in part; "large form of *Isospora bigemina"* of Mehlhorn, Heydorn, and Gestrich (1975) and Heydorn, Mehlhorn, and Gestrich (1975) in part; *Lucetina bigemina* (Stiles, 1891) Henry and Leblois, 1926; *Cryptosporidium vulpis* Wetzel, 1938 (probably); *Endorimospora miescheriana* (Kühn, 1865) Tadros and Laarman, 1976; *Sarcocystis bigemina* (Stiles, 1891) Levine, 1977.

Type Definitive Host. Domestic dog *Canis familiaris.*
Other Definitive Host. Wolf *C. lupus* and red fox *Vulpes vulpes.*
Intermediate Host. Pig *Sus scrofa.*

Location. Gamonts, gametes, zygotes, oocysts, and sporocysts in subepithelial tissue in the villi of the small intestine of the dog; sporocysts and oocysts have been found and only in the feces. The existence of more than 1 generation of meront (sarcocyst) has not been established. Sarcocysts in striated muscles of swine.

Geographic Distribution. Worldwide.

Prevalence. Common. Since all porcine sarcocysts were thought to be *S. miescheriana* until recently, it is impossible to give a definite figure for the prevalence of this species, but as high as 98.5% of pigs examined have been reported to be infected, and 44%–75% of the pigs in two surveys in the United States have been said to be positive (Levine, 1973). Farmer et al. (1978) said they found *Sarcocystis* in 36.5% of 123 sheep dogs from Wales and in 24% of 33 racing greyhounds from London. They identified *S. miescheriana* in 11 of 47 sheepdogs and in 2 of 8 greyhounds. Since the definitive host of another species whose sarcocysts occur in the pig *(S. hominis)* is the human, and since human feces are not easily available in pigs in the United States, it may be assumed that the figures given for this country are not far from correct. However, *S. porcifelis* has only recently been named from the pig in the USSR (see below), and it is not known whether it occurs in the United States.

The technique used for examination is important. Kozar (1971), for instance, found what she called *S. miescheriana* in 0.17%–0.53% of pigs in Poland by muscle section and in 2.4%–42% by trichinoscopy, but in 95% of 700 pigs using a new method involving grinding muscle in a mortar with a small amount of 0.85% NaCl solution and then looking for merozoites in the expressed fluid. Using tryptic digestion, Boch, Mannewitz, and Erber (1978) found *Sarcocystis* spp. in 35.5% of 409 older pigs and 9.7% of 74 fattened ones in south Germany; using the trichinoscope, they found them in only 5.6% and 3.9%, respectively. They found *S. miescheriana* in 48.3% of 122 *Sarcocystis*-positive slaughter pigs.

Oocyst Structure. The oocysts are similar in structure to those of *S. cruzi.* The sporulated oocysts seen by Stiles (1891) were said to be 13-16 x 8-10 μm. The sporocysts are 13 x 10 μm and apparently similar to those of *S. cruzi* (Rommel et al., 1974; Erber, 1978); 12-15 x 7-9 μm (Railliet and Lucet, 1891); 15 x 13 μm (Gassner, 1940); 13-16 μm long (Wenyon and Sheather, 1925) or 13-16 x 9-10 μm (Wenyon, 1926). They are 13-18 x 9-11 (mean, 15 x 10) μm in the red fox (Golemansky, 1975); they are without a Stieda body but have a residuum. The sporozoites are 7-8 x 3-4 μm in the red fox (Golemansky, 1975).

Sporulation. Occurs in the cells of the definitive host, but it has not been followed so far as we are aware. The feces contain free sporulated sporocysts or oocysts.

Merogony and Gametogony have not been described. The sarcocysts in porcine muscle are 0.5-4.0 mm long and up to 3 mm wide. They are compartmented, and their wall is striated with cytophaneres. Those described by Tadros and Laarman (1978) were up to 1.3 mm long and 350 μm in diameter, with vertically striated walls; in microisolated sarcocysts the striations are blunt, columnar protrusions 3.5 μm in average length. Mehlhorn, Hartley, and Heydorn (1976) found that the protrusions (cytophaneres) of the sarcocysts from a wild pig in Australia were 2.5-3.5 μm long and polygonal in cross section, with a maximum diameter of about 1.5 μm. In their interior were numerous fibrillar structures concentrated in a single layer of about 70 microtubules just beneath the primary cyst wall. There was no secondary cyst wall.

Prepatent Period. 8-9 days (Rommel et al., 1974).

Patent Period. Unknown.

Pathogenicity. Unknown for dogs (Dubey, 1976). Erber, Meyer, and Boch (1978) found that inoculation of pregnant sows caused abortion, fever, loss of condition, posterior weakness, and death.

Cross-Transmission Studies. This species is not transmissible from the pig to the cat (Rommel et al., 1974).

Remarks. This is the type species of the genus. Its history has been described briefly by Levine (1977).

So far as is known, the sporocysts of this species cannot be distinguished from those of *S. cruzi, S. tenella,* or *S. bertrami,* which also occur in the dog.

This is probably the species whose sporulated oocysts and sporocysts were found by Golemanski and Ridzhakov (1975) in the intestinal contents of 10% of 146 red foxes *V. vulpes* in Bulgaria. The oocysts were 17-20 x 10-18 (mean, 19 x 14) μm, the sporocysts 13-18 x 9-11 (mean, 15 x 10) μm, and the sporozoites 7-8 x 3-3.5 μm. In addition, Tadros and Laarman (1976) said that they had recovered "sporocysts identical to those shed by experimentally infected dogs . . . from intestinal scrapings of common red foxes in the immediate vicinity of the locality where wild swine with sarcocystic infection had been shot."

This is probably the organism that Erber (1978) found in the tongues of 97% of 96 wild boars in West Germany. He obtained sporocysts 13 x 10 μm in the feces of dogs, foxes, and a wolf, but not in various Felidae, after feeding them sarcocysts from the boars. The prepatent period was 10-12 days, and the patent period was 60 days.

This is also probably the organism that Wetzel (1938) found in red foxes in Germany and called *Cryptosporidium vulpis*. These fecal cysts were elongate, asymmetrically ovoid, 13-15 x 8-9 μm, with a colorless wall 0.5-μm thick, without a micropyle or Stieda body, and contained 4 sausage-shaped sporozoites with rounded ends, about 8 x 2.5 μm, plus a residuum. The sporozoites might possibly have contained a clear globule, but it was not definitely depicted.

This may also be the organism reported by Triffitt (1927) from a red fox *V. vulpes* bred for fur on Prince Edward Island, Canada, from the silver fox in Iowa by Lee (1934), and from the silver fox in Germany by Jacob (1949). Lee (1934) and Jacob (1949) said that they infected cats with cysts from the fox, but their claim is extremely dubious.

Levine, Cechner, and Meyer (1974) were unable to produce *S. miescheriana* sarcocysts in the muscles of baby pigs by feeding them oocysts of *Eimeria neodebliecki, E. debliecki, E. suis, E. scabra,* or *E. porci* from swine or *Isospora canis* from the dog.

Ashford (1977) thought that Stiles's *Coccidium bigemina* was a synonym of *S. tenella* of the sheep and dog.

Sarcocystis bertrami Doflein, 1901

Synonyms. "*Isospora rivolta*" sporocysts of Gassner (1940) and Levine and Ivens (1965a) in part; "large form of *Isospora bigemina*" of Mehlhorn, Heydorn, and Gestrich (1975) and Heydorn, Mehlhorn, and Gestrich (1975) in part; *Sarcocystis equicanis* Rommel and Geisel, 1975 (?); perhaps *Hoareosporidium pellerdyi* Pande, Bhatia, and Chauhan, 1972; *Endorimospora bertrami* (Doflein, 1901) Tadros and Laarman, 1976.

Type Definitive Host. Domestic dog *Canis familiaris* (?).

Intermediate Hosts. Domestic horse *Equus caballus*, domestic ass *Equus asinus*.

Location. Gamonts, gametes, zygotes, oocysts, and sporocysts are presumably in the subepithelial tissue in the villi of the small intestine of the dog, but, so far as we know, only sporocysts and oocysts have been found, and only in the feces. The existence of more than 1 generation of meront (sarcocyst) has not been established. Sarcocysts are in the striated muscles, and occasionally in the brain, of the horse and ass (Levine, 1973).

Geographic Distribution. Worldwide.

Prevalence. Common.

Oocyst Structure. The oocysts sporulate in the host. Presumably they are similar in structure to those of *S. cruzi*. The sporocysts of *S. equicanis* (see below) are 15–16 x 9–11 (mean, 15 x 10) μm (Rommel and Geisel, 1975).

Sporulation. Occurs in the cells of the definitive host, but it has not been followed, so far as we are aware. The feces contain free sporulated sporocysts.

Early *Merogony* and *Gametogony* have not been described. The sarcocysts in horse muscle are up to 10 mm long, are compartmented, and have a layer of cytophaneres on their outside.

Prepatent Period. Unknown.

Patent Period. Unknown.

Pathogenicity. Sarcocysts caused marked destruction of the muscles and chronic interstitial myositis in the horse.

Remarks. The sarcocysts that Doflein (1901) described for this species had a thick, radially striated wall like that of *S. miescheriana*. The wall described by Rommel and Geisel (1975) was much thinner, and that described by Göbel (1976) was intermediate. Tadros and Laarman (1976) remarked that if sarcocysts with thick and thin walls existed in horses, the specific name *bertrami* should be assigned to the thick-walled one and the name *equicanis* to the thin-walled one. The situation is more complicated than this, however.

Three different forms have been described:

1. *S. bertrami* Doflein, 1901. Doflein (1901) described the sarcocysts, presumably from Germany, as up to 9–10 mm long, compartmented, with a striated wall. He found them in the muscles and connective tissue, especially of the throat and front legs. They caused marked destruction of the muscles and chronic interstitial myositis. This was the organism seen by Siedamgrotzky (1872), with sarcocysts up to 12 x 0.3 mm and merozoites up to 16 x 5 μm.

2. *S. equicanis* Rommel and Geisel, 1975. These sarcocysts were found in 21 of 90 horses in Germany. They were up to 350 μm long, compartmented, with a very thin, unstriated wall less than 1 μm thick. They contained rounded metrocytes and slowly dividing elongate bradyzoites. No measurements were given for the merozoites, it being said merely that they were banana-shaped. Sporocysts were produced by feeding sarcocysts to dogs. They were 15–16 x 9–11 (mean, 15 x 10) μm. Göbel (1976) described the sarcocyst fine structure from the esophageal muscles of a horse. The sarcocysts were intracellular, with a wall about 1.5–2 μm thick. The primary layer of the wall was osmiophilic, about 56 nm thick, and apparently folded to form projections. It was septate and contained fibrils 8–12 nm thick. The merozoites in the compartments were piriform or spindle-shaped, 3–7 x 1–3 μm, with many white polysaccharide granules around the nucleus. They contained a conoid, a longitudinal mitochondrion, Golgi apparatus, ribosomes, rough endoplasmic reticulum, large numbers of rhoptries and micronemes, a micropore, and about 25 subpellicular microtubules. He saw endodyogeny.

3. *S. fayeri* Dubey et al., 1977. This species was found in 5 pools of 47–62 horses each by feeding dogs and in 4 esophagi and 2 hearts of 107 horses in Ohio. The sarcocysts were up to 900 x 70 μm, without obvious compartments, and had a striated wall 1–2 μm thick. The merozoites (bradyzoites) in the sarcocysts were banana-shaped, with one end pointed, and 15–20 x 2–3 μm when fixed and stained. Sporocysts were produced by feeding dogs. They were 11–13 x 7–9 (mean, 12 x 8) μm, and contained sporozoites 5–7 x 1–1.5 μm. In fixed sections, the sporocysts were 10.5–12.5 x 8.5 μm.

Rommel and Geisel (1975) considered their form different from *S. bertrami.* Rommel (in lit., 1977) wrote that he thought that *S. equicanis* was a synonym of *S. bigemina,* and *S. fayeri* possibly of *S. bertrami.*

Dubey et al. (1977) and Levine (1977) considered *S. equicanis* to be a synonym of *S. bertrami,* but we now wonder if this is true. Both Rommel and Geisel (1975) and Dubey et al. (1977) found that the dog (and not the cat) was the definitive host of their forms. What is puzzling is that Rommel and Geisel (1975) did not find the large sarcocysts of *S. bertrami* in what was apparently the same country in which they had been originally described, and Dubey et al. (1977) found only *S. fayeri* and neither *S. bertrami* nor *S. equicanis* in Ohio. Future research is needed to determine whether there are 1, 2, or 3 species of *Sarcocystis* in horses.

The written descriptions of all 3 species differ. The sarcocysts of *S. bertrami* differ from those of *S. equicanis* in being much larger, in having a striated wall, and in causing a tissue reaction; their sporocysts are unknown. Except for their striated wall, they differ similarly from those of *S. fayeri;* in addition, they are obviously compartmented, whereas the sarcocysts of *S. fayeri* are not. The sarcocysts of *S. equicanis* differ from those of *S. fayeri* in having a nonstriated wall and in having obvious compartments; in addition, their sporocysts are larger (15–16 x 9–11 μm vs. 11–13 x 7–9 μm). As to *S. bigemina,* its sarcocysts are unknown, but its sporocysts (12–15 x 7–9 μm according to Railliet and Lucet, 1891; dimensions not given by Stiles, 1891) are more the same size as those of *S. fayeri* (11–13 x 7–9 μm) than of *S. equicanis* (15–16 x 9–11 μm).

For the present, we think it best to list all 3 and to hope that future research will clarify the situation.

Sarcocystis equicanis Rommel and Geisel, 1975

Type Definitive Host. Domestic dog *Canis familiaris.*
Intermediate Host. Domestic horse *Equus caballus.*
Location. Oocysts and sporocysts in dog feces; sarcocysts in horse striated muscles.

Geographic Distribution. Europe (West Germany).
Prevalence. Found by Rommel and Geisel (1975) in 21 of 90 horses.
Oocyst Structure. See Remarks under *S. bertrami.*
Sporulation. Occurs in dog tissues.
Merogony and Gametogony. See Remarks under *S. bertrami.*
Prepatent Period. 8 days.
Patent Period. 3 weeks.
Pathogenicity. Apparently not pathogenic, either for the horse or dog.
Remarks. See Remarks under *S. bertrami.* This species may not be valid.

Sarcocystis fayeri Dubey et al. 1977

Type Definitive Host. Domestic dog *Canis familiaris.*
Intermediate Host. Domestic horse *Equus caballus.*
Location. Sporocysts are in the lamina propria near the tips of the villi of the small intestine of the dog, and gamonts, gametes, zygotes, and oocysts are also presumably there. So far as we know, only sporocysts and oocysts have been found in the feces.
Geographic Distribution. North America (Ohio).
Prevalence. Unknown.
Oocyst Structure. See Remarks under *S. bertrami.*
Sporulation. Occurs in the cells of the definitive host, but it has not been described, so far as we are aware. The feces contain free sporulated sporocysts.
Merogony and Gametogony. See Remarks under *S. bertrami.*
Prepatent Period. 11–14 days.
Patent Period. Unknown.
Pathogenicity. Apparently not pathogenic.
Remarks. See Remarks under *S. bertrami* above.

Sarcocystis levinei Dissanaike and Kan, 1978

Synonym. "Large form of *Isospora bigemina*" of Wenyon and Sheather (1925) and Wenyon (1926) in dogs from Sri Lanka.

Type Definitive Host. Domestic dog *Canis familiaris.*
Intermediate Host. Water buffalo *Bubalus bubalis.*
Location. Gamonts, gametes, zygotes, oocysts, and sporocysts in lamina propria of dog small intestine wall. Sarcocysts in water buffalo in striated muscles.
Geographic Distribution. Asia (Malaysia, Sri Lanka).
Prevalence. Common.
Oocyst Structure. Oocysts scraped from intestinal wall 21 x 15 μm, with smooth, thin, apparently 1-layered wall, without micropyle, residuum, or apparently polar granule. Wall of sporulated oocyst stretched between sporocysts. Sporocysts in dog feces 15–16 x 9–10 μm. Sporozoites in sections 8 x 1.5 μm.
Sporulation. Occurs in the small intestine wall of the definitive host.
Merogony. The number of generations of meronts in the intermediate host is unknown. The sarcocysts in the water buffalo are spindle-shaped, 0.8–1.15 x 0.09–0.14 (mean, 0.9 x 0.1) mm, with a thin, compartmented wall, and contain a few peripheral metrocytes and many banana-shaped merozoites 17–18 x 4–5 (mean, 18 x 4) μm. Under the electron microscope, the sarcocyst primary wall was 33–44 (mean, 39) nm thick, with invaginations 18–33 (mean, 26) nm deep, 56–89 (mean, 78) nm apart. The wall had sloping cytophaneres with an irregular, highly folded wavy wall 6–10 (mean, 7) μm high containing hollow, annulated, longitudinal fibrils and numerous coarse, electron-dense granules. Beneath the primary wall was a layer of ground substance about 2 μm thick that was also filled with fine fibrillar elements and coarse, electron-dense granules, and from which the sarcocyst compartments arose. The merozoites in the sarcocysts had 22 subpellicular microtubules, 200–300 micronemes, 8 rhoptries, conoid, nucleus, micropore, etc. There were also a few peripheral metrocytes, which had several micropores. Both merozoites and metrocytes had a 2-layered wall.
Remarks. This is most probably the "large form of *Isospora bigemina*" that Wenyon and Sheather (1925) and Wenyon (1926) found in dogs from Sri Lanka. It is not the *I. bigemina* (Stiles, 1891) Lühe, 1906 that Stiles (1891) found in dogs in France.

Sarcocystis capracanis Fischer, 1979

Type Definitive Host. Domestic dog *Canis familiaris.*
Type Intermediate Host. Domestic goat *Capra hircus.*
Location. Striated muscles of goat; small intestine and other organs of dog.
Geographic Distribution. Europe (Germany); presumably worldwide.
Oocyst Structure. Oocysts not described. Sporocysts 12–15 x 9–10 (mean, 14 x 9.5) μm.
Sporulation. Occurs in small intestine of definitive host.
Merogony. Fischer (1979) found first-generation meronts in all the organs he studied in the goat (heart, liver, spleen, lungs, brain, and mesenteric lymph nodes) 10–12 days after inoculation. They were in the endothelial cells of the small blood vessels. He saw only single merozoites 10 days after inoculation, and immature meronts and many merozoites 12 days after inoculation. The merozoites were 6–8 x 2–4 μm.

Fischer (1979) found second-generation meronts in the same locations 20–23 days after inoculation.

Fischer (1979) found sarcocysts containing metrocytes only in the heart, muscles, striated muscles, and brain of the goat on day 43 after inoculation. Merozoites appeared on day 65. Unfixed, unstained merozoites were about 14 x 4 μm, and merozoites in fixed, stained smears 12–16 x 3–5 μm. On day 92 the sarcocysts were 130–800 x 50–70 μm. Sarcocysts in naturally infected goats were up to 3 mm long. The sarcocyst wall averaged 2.8 μm thick and appeared striated. Unfixed sarcocysts had finger-formed processes in their walls.

Gametogony. Occurs in the small intestine of the dog.
Prepatent Period. 7–8 days.
Patent Period. Several weeks.
Pathogenicity. This species is not pathogenic for the dog, but Fischer (1979) found that the first- and second-generation meronts were pathogenic for the goat. Inoculation of 50,000 to 2 million sporocysts by mouth caused all goats to become ill within 20 days. Characteristic clinical signs included very high body temperature and anemia. Two goats fed 2 million sporocysts each and one each of those fed 100,000, 80,000 and 50,000, respectively, died 20–65

days after inoculation. Two goats fed 80,000 and 100,000 sporocysts, respectively, were killed on days 31 and 43 while moribund.

Cross-Transmission Studies. Fischer (1979) could not infect 2 sheep of this species.

Remarks. The life cycle of this species was worked out by Fischer (1979) in Germany. D. K. Pethkar (personal communication) in India found 2 types of microscopic sarcocyst in goats. One had a radially striated wall 2–4 μm thick, and the other had a thin, smooth wall with hairlike processes about 8 μm long. He, too, transmitted this (or these?) organism to the dog. *S. capracanis* has microscopic sarcocysts and may be Pethkar's first type. *S. capracanis* is apparently not *S. moulei* Neveu-Lemaire, 1912 from the domestic goat, since the latter has large sarcocysts, but its relationship to *S. orientalis* Machul'skii and Miskaryan, 1958 from the wild goat *Capra sibirica* in the USSR is unknown; Fischer (1979) failed to mention this species. The number of species in the goat remains to be determined.

Sarcocystis horvathi Ratz, 1908

Synonyms. *Sarcocystis gallinarum* Krause and Goranoff, 1933.
Type Definitive Host. Domestic dog *Canis familiaris.*
Other Definitive Host. Perhaps domestic cat *Felis catus.*
Type Intermediate Host. Domestic chicken *Gallus gallus.*
Location. Striated muscles.
Geographic Distribution. Europe (Hungary), USSR, Australia, Oceania (New Guinea).
Prevalence. Munday, Humphrey, and Kila (1977) found this species in 4% of 78 chickens in Papua, New Guinea, and 2 chickens in Australia.
Oocyst Structure. Oocysts in jejunum about 15 x 11 μm; sporocysts in feces 10–13 x 7–9 (mean, 11.5 x 8) μm, with 4 sporozoites and residuum (Munday, Humphrey, and Kila, 1977).
Sporulation. Occurs in intestine of definitive host.
Merogony. Occurs in chicken muscles. With striated wall (Munday, Humphrey, and Kila, 1977).
Gametogony. Occurs in intestine of definitive host.

Prepatent Period. Munday, Humphrey, and Kila (1977) found that the prepatent period in the dog was 6 days. Golubkov (1979) said that the prepatent period in the dog was 10 days.

Patent Period. More than 4 days (Munday, Humphrey, and Kila, 1977, killed the host dog at day 4). Golubkov (1979) said that the patent period in the dog was 21–23 days.

Pathogenicity. Munday, Humphrey, and Kila (1977) said that this organism caused a severe myositis in the 5 infected chickens they examined.

Remarks. Golubkov (1979), who found sarcocysts in the chicken in the USSR, said that he transmitted them to both the dog and cat. The prepatent period in the cat was apparently 11 days, and the patent period was apparently 11–19 days.

Sarcocystis rileyi (Stiles, 1893) Minchin, 1903

Golubkov (1979) said that he found sarcocysts of *Sarcocystis horvathi* in the domestic chicken and *S. rileyi* in the domestic duck in the USSR and produced sporocysts in dogs and cats by feeding them sarcocysts from both species of birds. The prepatent period (apparently for both) was 10 days in the dog and 11 days in the cat. The patent period (apparently for both) was 21–23 days in the dog and 11–19 days in the cat. The sporocysts in the dog were 13.5–15.4 x 10.6 ± 0.5 μm, and the sporozoites were 9.3 x 2.8 μm. He gave no more data. It would certainly be unusual if both the dog and cat were definitive hosts of the same species.

Sarcocystis spp. Janitschke, Protz, and Werner, 1976

Janitschke, Protz, and Werner (1976) found sarcocysts of this form in Grant's gazelle *Gazella granti* in Tanzania and obtained sporocysts and sporulated oocysts by feeding the meat to cats and a dog. The sporocysts in the cat were 11–16 x 8–12 (mean, 13 x 9) μm, while those in the dog were 13–18 x 8–12 (mean, 16 x 11) μm. The prepatent period in the cats was 20 days; that in the dog was 10 days. The patent period in the cats was 44–48 days; that in the dog was 42 days. They thought that they were dealing with two species, but in the absence of further infection studies they did not know which was which.

Three species of *Sarcocystis* have been named from *Gazella*: *S. gazellae* Balfour, 1913 was found in *G. rufifrons* in the Sudan, *S. woodhousei* Dogel', 1916 in *G. granti* in East Africa, and *S. mongolica* Machul'skii, 1947 in *G. gutturosa* in Buryat-Mongolia. Janitschke, Protz, and Werner (1976) mentioned the first 2. In addition, *Sarcocystis* spp. have been reported from *G. thomsoni*, *G. subgutturosa*, and *Gazella* sp. by various authors (see Kalyakin and Zasukhin, 1975). What names to assign to these species remains to be determined.

Sarcocystis hemionilatrantis Hudkins and Kistner, 1977

Type Definitive Host. Coyote *Canis latrans.*

Other Definitive Host. Domestic dog *C. familiaris.*

Intermediate Host. Mule deer *Odocoileus hemionus hemionus.* Hudkins and Kistner (1977) were unable to infect a calf *Bos taurus* or 2 lambs *Ovis aries* with sporocysts from the coyote.

Location. Gamonts, gametes, zygotes, oocysts, and sporocysts are presumably in the subepithelial tissue in the villi of the small intestine of the coyote and dog, but, so far as we know, only sporocysts and oocysts have been found, and only in the feces. Meronts are in striated muscles and macrophages of mule deer.

Geographic Distribution. North America (Oregon).

Oocyst Structure. Oocysts presumably similar in structure to those of *S. cruzi.* Sporocysts 14–16 x 9–12 (mean, 14 x 9) μm, with a residuum (Hudkins-Vivion, Kistner, and Fayer, 1976; Hudkins and Kistner, 1977).

Sporulation. Occurs in the cells of the definitive host, but it has not been described, so far as we are aware. The feces contain free sporulated sporocysts.

Merogony and Gametogony. There appear to be 2 types of meronts in mule deer. In fawns that died 27–39 days after ingesting sporocysts from the coyote, Hudkins and Kistner (1977) found microscopic meronts in macrophages, between muscle fibers and near blood vessels in the esophagus, heart, biceps, femoris, semimembranosus, diaphragm, and tongue. They did not find sarcocysts in the muscles until 60 days after inoculation; their size was not given.

Prepatent Period. 9–12 days or more (Hudkins-Vivion, Kistner, and Fayer, 1976; Hudkins and Kistner, 1977).

Patent Period. Uncertain. Coyote pups shed sporocysts in their feces intermittently 12–36 days after having been fed infected mule deer meat (Hudkins and Kistner, 1977).

Pathogenicity. Apparently not pathogenic for the coyote or dog. However, Hudkins and Kistner (1977) reported that 9 of 11 mule deer fawns fed sporocysts from the coyote died 27–63 days later with clinical signs of anorexia, weight loss, fever, and weakness.

Remarks. The relationship of this species to *S. gracilis* Ratz, 1908 (described from the red deer *Cervus elaphus*), *S. grueneri* Yakimoff and Sokoloff, 1934 (described from *Cervus elaphus sibiricus*), *S. cervi* Destombes, 1957 (described from an unidentified species of deer in Vietnam), and to other species of *Sarcocystis* reported from deer (see Kalyakin and Zasukhin, 1975) remains to be determined. This may be the same species that Honess (1956) found in the mule deer *O. hemionus* and elk *Cervus canadensis* in Wyoming, that Karstad and Trainer (1969) found in sections of the tongue muscles in 79% of 209 white-tailed deer *O. virginianus* from Ontario, Wisconsin, and Texas, that Prestwood, Pursglove, and Hayes (1976) found in white-tailed deer in West Virginia, and that Blažek, Kotrlý, and Ippen (1976) found in a "Virginia" deer in Czechoslovakia. It may also be the same species (with sporocysts 14–17 x 8–11 [mean, 16 x 10] μm) that Arther and Post (1977) found in 21% of 82 coyotes in Colorado.

Sarcocystis sp. Erber, 1978

Erber (1978) found 3 types of sarcocysts in the tongues or abdominal musculature of 93% of 421 roe deer *Capreolus capreolus* in West Germany: Type 1. Sarcocyst wall smooth, about 1 μm thick; Type 2. Sarcocyst wall with fine, hairlike cytophaneres 6–8 μm long and much less than 0.5 μm thick; Type 3. Sarcocyst wall with rigid, fingerlike cytophaneres 5–6 μm long and about 0.5 μm thick.

He fed different Canidae and Felidae raw muscles infected with Type 1 and Type 2 sarcocysts. Dogs and foxes but apparently not Felidae shed sporocysts 16-18 x 9-12 μm in the feces. Their prepatent period was 10-13 days and their patent period 50 days.

Sarcocystis spp. Entzeroth, Scholtyseck, and Greuel, 1978

Entzeroth, Scholtyseck, and Greuel (1978) fed esophagus, diaphragm, and peritoneal muscles of roe deer *Capreolus capreolus* to a fox and a dog. After a prepatent period of 8 days, they found sporulated oocysts 14.5 x 8.5 μm in the fox. After a prepatent period of 10-14 days, they found sporulated oocysts 15.6 x 10.0 μm in the dog; the patent period in the dog was 51 days.

Sarcocystis sp. Sahasrabudhe and Shah, 1966

Sarcocysts of this organism were found in the muscles of an esophageal nodule of a dog in Madhya Pradesh, India. They were spherical to elongate, 110-250 μm in diameter, and contained thousands of crescent-shaped merozoites 4-5 x 1.5 μm. The sarcocysts had no septa. Their wall was hyaline and 1-2 μm thick. This may have been a species found normally in the muscles of some prey animal.

Toxoplasma (?) sp. (Gill et al. 1978)

Synonym. Hammondia sp. or *Isospora* sp. of Gill et al., 1978.
Definitive Host. Domestic dog *Canis familiaris.*
Intermediate Host. Water buffalo *Bubalus bubalis.*
Location. Feces of dog; diaphragm of water buffalo.
Geographic Distribution. Asia (India).
Oocyst Structure. Spherical or subspherical, 17-24 x 16-19 (mean, 13.3 *[sic]* x 18.2) μm, with smooth, yellowish, 2-layered wall 1-2 μm thick, without micropyle, residuum, or polar granule. Sporocysts ellipsoidal, 11-16 x 9-11 (mean, 13 x 10) μm, without Stieda body, with residuum composed of numerous small scattered

granules. Free sporozoites banana-shaped with blunt ends but very
rarely with one end pointed, with central or subcentral nucleus,
with 2 clear globules, one on each side of nucleus, 7–10 x 2–4
(mean, 8 x 3) μm.

Sporulation. 8–16 hours at room temperature (in India) in 2.5%
potassium bichromate solution.

Prepatent Period. 8–9 days.

Patent Period. 15–25 (mean, 20) days.

Remarks. Unsporulated oocysts of this species were found in 4
dogs fed diaphragm muscles from water buffalo naturally infected
with macroscopic sarcocysts of (presumably) *Sarcocystis fusiformis;*
the authors did not look for microscopic sarcocysts. They thought
that the oocysts were not those of *Sarcocystis* since they were unsporulated and thought it impossible to decide whether they were
those of *"Hammondia"* or *Isospora.* They are about the same size
as the oocysts of *I. burrowsi* and *I. ohioensis.*

Besnoitia (?) sp. (Shelton, Kintner, and MacKintosh, 1968)

Synonym. Coccidium-like organism of Shelton, Kintner, and
MacKintosh, 1968.

Definitive Host. Unknown.

Intermediate Host. Domestic dog *Canis familiaris.*

Location. Skin and subcutaneous tissues, mostly in large macrophages.

Geographic Distribution. North America (Missouri).

Oocyst Structure. Oocysts unknown.

Merogony. Subcutaneous meronts about 5 mm in diameter,
containing merozoites about 5–6 x 1–2 μm.

Remarks. Shelton, Kintner, and MacKintosh (1968) found this
organism in a dog in Missouri. It appears to be more like *Besnoitia*
than like any other genus.

Isospora dutoiti Yakimoff, Matikaschwili, and Rastegaieff, 1933

(Plate 15, Fig. 81)

Synonym. Eimeria dutoiti Yakimoff, Matikaschwili, and Rastegaieff, 1933 *lapsus calami.*

Type Host. Golden jackal *Canis aureus.*
Location. Unknown, oocysts found in feces.
Geographic Distribution. USSR (Transcaucasia).
Oocyst Structure. Oocysts spherical and sometimes slightly ellipsoidal (described as oval), the ellipsoidal forms being 10–13 x 9–11 (mean, 11.5 x 10) μm and the spherical forms 9–13 μm in diameter (mean, 11 μm), illustrated with a 1-layered wall, without micropyle, residuum, or polar granule. Sporocysts spherical, 7 μm in diameter, without Stieda body, with residuum. Sporozoites piriform, 6–7 x 2 μm.

Remarks. Pellérdy (1974) said that the scanty data presented in the description scarcely allow the determination of this species. Perhaps someone will rediscover and redescribe it in the future. In the meantime, its validity is suspect.

Isospora theileri Yakimoff and Lewkowitsch, 1932

(Plate 15, Fig. 83)

Type Host. Golden jackal *Canis aureus.*
Location. Unknown. Oocysts found in feces.
Geographic Distribution. USSR (Azerbaidzhan).
Oocyst Structure. Oocysts spherical or slightly ovoid to ellipsoidal, 21 x 17–18 μm, illustrated with a 1-layered wall, without micropyle or residuum. Sporocysts 13–16 x 9–11 (mean *[sic]*, 16 x 11) μm, illustrated without Stieda body, with residuum.

Remarks. Yakimoff and Lewkowitsch (1932) failed to transmit this species to the domestic dog by feeding sporulated oocysts. It, too, should be restudied if it is ever rediscovered. During transportation to Leningrad almost all the oocyst walls disappeared, freeing the sporocysts.

Sarcocystis tropicalis (Mukherjea and Krassner, 1965)
Levine and Tadros, 1980

Synonym. Isospora tropicalis Mukherjea and Krassner, 1965.
Type Definitive Host. Golden jackal *Canis aureus.*
Intermediate Host. Unknown.

Location. Subepithelial tissues of small intestine and also apparently villar epithelial cells according to their photomicrograph.
Geographic Distribution. Asia (India).
Oocyst Structure. Oocysts said to be 16 μm in diameter but illustrated as longer in one direction than the other. No oocysts were found in the feces, although sporulated ones without micropyle or residuum, and with a thin wall stretched around 2 sporocysts, were seen in the intestinal contents. Free sporocysts in feces ellipsoidal, 15–16 x 10–12 μm, with a relatively thick wall, with a large residuum, with 4 banana-shaped sporozoites, each 8–10 x 3–4 μm.
Sporulation. Occurred in the host small intestine.
Cross-Transmission Studies. Mukherjea and Krassner (1965) fed sporulated oocysts to a "clean" young Indian fox *Vulpes bengalensis,* which died 12 days later. They found no oocysts or sporocysts in its feces 3, 5, and 7 days after inoculation, but found many sporocysts in the contents of the small intestine 12 days after inoculation.
Remarks. Pellérdy (1974) said that it was problematic whether the sporocysts they found in the fox intestine were in fact those of this species, and we agree.
It is obvious that this is a *Sarcocystis.*

Host Genus *Alopex*

Eimeria mesnili Rastegaieff, 1929

Type Host. Arctic or blue fox *Alopex lagopus.*
Location. Unknown; oocysts found in feces.
Geographic Distribution. USSR (Murman).
Oocyst Structure. Oocysts spherical to ovoid, 18 x 11–14 μm, with wall illustrated as composed of 1 layer, with micropyle occupying whole small end, without residuum. Sporocysts appear ellipsoidal in the illustration, but the drawing is poor.
Remarks. Pellérdy (1974) spoke of this as a "species (?)." The oocyst looks suspiciously like a rabbit coccidium. However, Rastegaieff (1929, 1930) said that she had found it in the same fox on 10 October and 19 November 1928.

Eimeria imantauica Nukerbaeva and Svanbaev, 1973

Type Host. Arctic fox *Alopex lagopus.*
Location. Intestine.
Geographic Distribution. USSR (Kazakhstan).
Prevalence. Nukerbaeva and Svanbaev (1973) found this species in 2.7% of 1,089 Arctic foxes in Kazakhstan.
Oocyst Structure. Ellipsoidal, 13–15 x 8–11 (mean, 14 x 10) μm, with 2-layered wall 1 μm thick, without micropyle or polar granule, with residuum; sporocysts ellipsoidal ("oval"), 6–7 x 3–4 μm, without Stieda body or residuum; sporozoites elongate, with one end rounded and the other pointed, lying lengthwise head to tail in sporocysts, with clear globule at the broad end.

Isospora pavlodarica Nukerbaeva and Svanbaev, 1973

See under *Vulpes vulpes.*

Isospora triffitti Nukerbaeva and Svanbaev, 1973

See under *Vulpes vulpes.*

Host Genus *Vulpes*

Eimeria adleri Yakimoff and Gousseff, 1936

(Plate 2, Fig. 8)

Type Host. Red fox *Vulpes vulpes;* silver fox *V. vulpes.*
Geographic Distribution. USSR (Western USSR, Kazakhstan).
Oocyst Structure. Oocysts spherical, colorless, 18–25 μm in diameter, with a smooth, yellow-green, 1-layered wall 1.2–1.4 μm thick, without micropyle, residuum, or polar granule. Sporocysts broadly ovoid to ellipsoidal, 9–16 x 8–10 μm, without Stieda body or residuum. Sporozoites comma-shaped, 6–11 x 4–5 (mean, 9 x 4) μm, lying lengthwise head to tail in sporocysts, with clear globule

μm, lying lengthwise head to tail in sporocysts, with clear globule at broad end (Yakimoff and Gousseff, 1936; Svanbaev, 1960).

Sporulation. 4 days at 25 C in 2% potassium bichromate solution (Svanbaev, 1960).

Remarks. This species was also reported (without description) from the fox in Byelorussia by Litvenkova (1969).

Eimeria vulpis Galli-Valerio, 1929

Type Host. Red fox *Vulpes vulpes;* silver fox *V. vulpes.*
Location. Unknown; oocysts found in feces.
Geographic Distribution. Europe (England, Switzerland, Bulgaria), USSR (Kazakhstan, Transcaucasia).
Prevalence. Watkins and Harvey (1942) found this species (the only coccidium they saw) in 11 out of 52 silver fox cubs dying out of pelting season in England. They also found it in perhaps 10% of the adult and perhaps 25% of the cub foxes from some 15 fox farms in England. Svanbaev (1960) found it in 22% of 18 silver foxes in the Alma Atinsk Oblast, Kazakhstan. Golemanski and Ridzhakov (1975) found its oocysts in the feces of 10% of 146 red foxes in Bulgaria. We have not seen it in the United States.
Oocyst Structure. Ovoid or ellipsoidal, 16–26 x 12–24 (mean, 17–22 x 14–16) μm, with smooth, colorless wall 0.8–1.5 μm thick, with barely visible micropyle or without one, apparently without residuum, without polar granule. Sporocysts ovoid to ellipsoidal, 5–6 x 3–6 (mean, 6 x 4–5) μm, without Stieda body, with residuum. Sporozoites comma-shaped, 4–5 x 2 μm.
Sporulation. 3–4 days at 25 C in 2% potassium bichromate solution (Svanbaev, 1960).

Eimeria li Golemansky, 1975

(Plate 4, Fig. 23)

Type Host. Red fox *Vulpes vulpes.*
Location. Unknown; oocysts found in feces or large intestine contents.
Geographic Distribution. Europe (Bulgaria).

Prevalence. Golemansky (1975) found this species in 1.5% of 146 red foxes in Bulgaria. Both foxes were 1 month old.

Oocyst Structure. Ovoid, 29–33 x 22–25 (mean, 31 x 23) μm, colorless to slightly yellowish, with 2-layered wall 2.5 μm thick, the outer layer the lighter and thicker one, without micropyle or residuum, with 2–6 polar granules. Sporocysts elongate ovoid, 14–18 x 6–10 (mean, 16 x 8) μm, without Stieda body, with residuum. Sporozoites elongate, 12 x 3 μm, lying lengthwise head to tail in sporocysts, with clear globule at large end.

Sporulation. 86 hours at 23 C in 3% potassium bichromate solution.

Eimeria sp. Golemanski and Ridzhakov, 1975

Type Host. Red fox *Vulpes vulpes.*
Location. Unknown; oocysts found in feces.
Geographic Distribution. Europe (Bulgaria).
Prevalence. 4% of 146 red foxes in Bulgaria (Golemanski and Ridzhakov, 1975).
Oocyst Structure. Oocysts 28–38 x 17–23 (mean, 33 x 20) μm. Sporocysts 13–15 x 8–10 μm, without Stieda body.
Remarks. Golemanski and Ridzhakov (1975) considered this form a pseudoparasite of the fox and probably a rabbit species.

Eimeria heissini Svanbaev, 1956

(Plate 4, Fig. 20)

Type Host. Corsac fox *Vulpes corsac.*
Location. Unknown; oocysts found in feces.
Geographic Distribution. USSR (Kazakhstan).
Oocyst Structure. Spherical, 20 μm in mean diameter, with smooth, colorless, double-contoured wall 1 μm thick, without micropyle or polar granule, with residuum. Sporocysts spherical, 8 μm in mean diameter, without Stieda body, with residuum. Sporozoites ovoid, 5 x 3 μm, apparently without clear globule.
Sporulation. 3–4 days at 20–25 C in potassium bichromate solution (Svanbaev, 1956).

Eimeria lomarii Dubey, 1963

(Plate 4, Fig. 24)

Type Host. Indian fox *Vulpes bengalensis.*
Location. Unknown. Oocysts found in intestine.
Geographic Distribution. Asia (India).
Oocyst Structure. Ellipsoidal, 24–29 x 19–22 μm, with 2-layered
wall 1–2 μm thick, outer layer relatively thick, grayish, inner layer
darker, without micropyle, residuum, or polar granule. Sporocysts
broadly ovoid, 11–14 x 8–10 μm, with vestigial Stieda body and
residuum. Sporozoites comma-shaped, 8–10 x 2–3 μm, with clear
globule at broad end.
Sporulation. 4 days.

Isospora vulpis Galli-Valerio, 1931

Type Host. Red fox *Vulpes vulpes.*
Location. Unknown. Oocysts found in feces.
Geographic Distribution. Europe (Switzerland, Bulgaria) USSR
(Transcaucasia).
Prevalence. Golemanski and Ridzhakov (1975) found this species
in the feces of 8% of 146 red foxes in Bulgaria.
Oocyst Structure. Galli-Valerio (1931) said that the oocysts were
ovoid, 25 x 24 μm, with a micropyle, and that the sporocyst did
not have a residuum. He gave no more information. Golemanski
and Ridzhakov (1975) said that the oocysts they found were spherical with a diameter of 18–26 μm or oval and 20–23 x 18–21 μm,
and had sporocysts, 16–18 x 8–11 μm.

Isospora (?) sp. (Ashford, 1977) nov. comb.

Synonym. Hammondia sp. Ashford, 1977.
Type Host. Red fox *Vulpes vulpes.*
Location. Unknown; oocysts found in feces.
Geographic Distribution. Europe (England).

Prevalence. Ashford (1977) found this form in the feces of 1 of 22 naturally infected red foxes in England.

Oocyst Structure. Oocysts 14 x 12 µm. No other information given.

Sporulation. 2–3 days in bichromate solution, presumably at room temperature.

Remarks. Ashford (1977) said that these oocysts resembled those of *"Hammondia" hammondi,* but were not infective for mice. He thought them to be another species of *"Hammondia,"* but they could just as well have been *Isospora* or perhaps *Besnoitia.*

Isospora canivelocis Weidman, 1915 emend. Wenyon, 1923

(Plate 5, Figs. 27, 29, 30)

Synonyms. Coccidium bigeminum var. *canivelocis* Weidman, 1915; *Isospora bigemina* var. *canivelocis* (Weidman, 1915) Mesnil, 1916; *I. canivecolis [sic]* (Weidman, 1915) Wenyon, 1923; *I. canivelocis* (Weidman, 1915) Wenyon, 1923 of *auctores; Lucetina canivelocis* (Weidman, 1915) Henry and Leblois, 1926.

Type Host. Swift fox *Vulpes velox.*

Other Hosts. Silver fox *V. vulpes,* red and silver foxes *V. vulpes,* Arctic fox *Alopex lagopus* (experimental).

Location. Unknown; oocysts found in feces.

Geographic Distribution. North America (Philadelphia Zoo, Iowa), Europe (Bulgaria, Germany, Holland, Hungary), USSR (Kazakhstan).

Prevalence. Svanbaev (1960) reported this species in 33% of 18 silver foxes in the Alma Atinsk Oblast, Kazakhstan. Golemanski and Ridzhakov (1975) found it in 15% of 146 wild red foxes in Bulgaria. Nukerbaeva and Svanbaev (1973) found it in 11.5% of 1,199 silver foxes in Kazakhstan.

Oocyst Structure. Sporulated oocysts subspherical to ellipsoidal, 27–38 x 25–30 (mean, 33 x 28) µm according to Weidman (1915), 24–32 x 20–32 (mean, 27 x 26) µm according to Svanbaev (1960), 33–39 x 27–32 (mean, 36 x 30) µm according to Nieschulz and Bos (1933), 21–38 x 17–31 µm according to Sprehn and Cramer

(1931) or 30–39 x 23–30 (mean, 35 x 27) μm according to Golemanski and Ridzhakov (1975), with smooth, yellowish orange or
yellowish brown, 1-layered wall 1.2–1.6 μm thick, with oocyst wall
in old oocysts sometimes collapsed around sporocysts, without
micropyle, residuum, or polar granule. Sporocysts ellipsoidal, 14–
16 μm in diameter according to Weidman (1915), 14–24 x 11–14
μm according to Sprehn and Cramer (1931), 16–20 x 9–12 (mean,
18 x 11) μm according to Svanbaev (1960) or 11–17 x 9–11 μm
according to Golemanski and Ridzhakov (1975), without Stieda
body, with residuum. Sporozoites banana- or comma-shaped, 10–12
x 3–4 (mean, 11 x 4) μm according to Svanbaev (1960) or 12–15
x 3–4 μm according to Golemanski and Ridzhakov (1975). The
oocysts seen by Nukerbaeva and Svanbaev (1973) in silver foxes
were "oval," 28–31 x 25–28 (mean, 29 x 26) μm, with a 2-layered
wall 1.5 μm thick, inner layer yellow-orange, without micropyle
or polar granule; sporocysts 14–20 x 10–13 μm, with residuum.

Sporulation. Apparently 1 day (Weidman, 1915) or 2–4 days
at 25 C in 2% potassium bichromate solution (Svanbaev, 1960).

Cross-Transmission Studies. No attempts have been made to
infect other species of animals with oocysts from the swift fox.
Using material from the silver fox, Sprehn and Cramer (1931)
were unable to infect 2 kittens, while Pellérdy (1955) could not
infect a badger. Nieschulz and Bos (1933) were unable to transmit
the larger form (see below) to a young dog of unknown history.
Nukerbaeva and Svanbaev (1974) infected *A. lagopus* with what
they called this species from *V. vulpes* and vice versa, but they
were unable to infect the cat, dog, or mink with oocysts from *V.
vulpes.*

Remarks. Pellérdy (1974) gave the host of this species as *V.
vulpes,* but the type host is the swift fox *V. velox,* not *V. vulpes.*

As originally described, the oocysts were said to be 27–38 μm
long (Weidman, 1915). However, Sprehn and Cramer (1931) found
what they considered to be this species in silver foxes in Germany
or Poland. The oocysts of their form were 21–38 μm long, but had
2 modes. The smaller and larger oocysts were somewhat different,
the larger ones having a point at one end and the smaller ones being
broadly "ovoid" without tapering at the poles. Nieschulz and Bos
(1933) also found what they considered to be 2 sizes of this species
in young silver foxes on a fox farm in the Netherlands. They re-

tained the name *I. canivelocis* for the large form with oocysts 33–
39 μm long with one pole clearly pointed, and named the small
form *I. vulpina*. As indicated below, the oocysts described as those
of *I. vulpina* from silver foxes in Wisconsin by Bledsoe (1976) were
intermediate between the size ranges given by Nieschulz and Bos
(1933) for *I. vulpina* and *I. canivelocis;* they were clearly ellipsoidal,
however, and did not have a point at one end. The fact that the
only report of *I. canivelocis* from the swift fox is that of Weidman
(1915) raises the question as to what would be discovered if more
swift foxes were examined.

Isospora buriatica Yakimoff and Machul'skii, 1940

(Plate 10, Fig. 55)

Type Host. Corsac fox *Vulpes corsac.*
Other Hosts. Indian fox *V. bengalensis,* silver fox *V. vulpes.*
Location. Unknown; oocysts found in feces.
Geographic Distribution. USSR (Tashkent, Kazakhstan), Asia
(India).
Prevalence. Nukerbaeva and Svanbaev (1973) said they found
this species in 12.4% of 1,199 silver foxes in Kazakhstan.
Oocyst Structure. The following description of the oocyst is
from Svanbaev (1956), who found this species in the corsac in
Kazakhstan; we have not seen the original description by Yakimoff
and Machul'skii (1940) from the corsac in the Tashkent zoo. Oocysts ellipsoidal, 25 x 19 μm, or ovoid, 32 x 25 μm, with smooth,
double-contoured, greenish or yellowish green wall 1–1.3 μm thick,
without micropyle or residuum, sometimes with polar granule.
Sporocysts ovoid, 15–17 x 10–13 μm, with residuum. Sporozoites
comma-shaped, 6–8 x 3–4 μm. The following description is from
Dubey (1963), who reported this species in the Indian fox. Oocysts
ovoid, 29–33 x 22–25 (mean, 31 x 25) μm, with 2-layered wall
1–2 μm thick, outer layer relatively thick and yellowish, inner
darker, without micropyle, residuum, or polar granule. Sporocysts
ellipsoidal, 19–21 x 11–12 (mean, 19 x 12) μm, with a 2-layered
wall, without Stieda body, with residuum. Sporozoites 14–15 x
3–4 μm, with clear globule at broad end. The oocysts described

by Nukerbaeva and Svanbaev (1973) from the silver fox in Kazakhstan were ovoid, 31–45 x 25–31 (mean, 36 x 28) μm, with a 2-
layered wall 1.7–2.5 μm thick, without micropyle or polar granule;
sporocysts "oval," 20–22 x 11–14 μm; sporozoites, 10–13 x 4.6
μm.

Sporulation. 3–4 days at 20–25 C in potassium bichromate solution (Svanbaev, 1956), 1–2 days (Nukerbaeva and Svanbaev, 1973),
or 2 days (Dubey, 1963).

Remarks. Whether the differences among the above descriptions
are significant remains to be seen.

Isospora vulpina Nieschulz and Bos, 1933

(Plate 5, Figs. 25, 26)

Synonyms. Isospora vulpina var. *vulpina* Mantovani, 1965; *Iso-
spora vulpina* var. *aprutina* Mantovani, 1965; *Isospora aprutina*
Mantovani, 1965 emend. Pellérdy, 1974; *Cystoisospora vulpina*
(Nieschulz and Bos, 1933) Frenkel, 1977.

Type Host. Silver fox *Vulpes vulpes,* red fox *V. vulpes,* Arctic
fox *Alopex lagopus.*

Other Hosts. Domestic dog *Canis familiaris* (experimental).

Transport Host. House mouse *Mus musculus.*

Location. Unknown; oocysts found in feces or contents of large
intestine.

Geographic Distribution. Europe (Holland, Italy, Bulgaria), USSR
(Georgia, Kazakhstan), North America (Washington, Wisconsin).

Prevalence. Common. Golemanski and Ridzhakov (1975) found
this species in 18% of 146 red foxes in Bulgaria. It was the only
species that Bledsoe (1976) found in silver foxes in Wisconsin.
Mikeladze (1978) found it in foxes in Georgia, USSR. Nukerbaeva
and Svanbaev (1973) found it in 5.2% of 1,199 silver foxes in
Kazakhstan.

Oocyst Structure. Oocysts broadly ellipsoidal, 21–32 x 19–27
(mean, 25 x 21) μm according to Nieschulz and Bos (1933), 22–28
x 18–22 (mean, 24.5 x 21) μm according to Mantovani (1965),
25–38 x 21–32 (mean, 30, x 24) μm according to Bledsoe (1976),
21–34 x 19–28 (mean, 27 x 23) μm according to Dunlap (1956)

or 20–31 x 16–24 (mean, 25 x 20) μm according to Golemanski
and Ridzhakov (1975), with smooth, yellowish, 1-layered wall
about 1 μm thick, sometimes with wall collapsed around sporocysts after sporulation, without micropyle, residuum, or polar
granule. Sporocysts ellipsoidal to slightly ovoid, 13–18 x 9–13
(mean, 15 x 11) μm according to Nieschulz and Bos (1933), 17–
20 x 12–14 (mean, 18.5 x 13) μm according to Mantovani (1965),
15–23 x 11–16 (mean, 18 x 13) μm according to Bledsoe (1976)
or 16–19 x 10–13 μm according to Golemanski and Ridzhakov
(1975), without Stieda body, with residuum. Sporozoites elongate,
10–12 x 2–3 μm according to Golemanski and Ridzhakov (1975),
lying lengthwise in sporocysts, without clear globule. The oocysts
seen by Nukerbaeva and Svanbaev (1973) in silver foxes in Kazakhstan were subspherical, 22–28 x 17–22 (mean, 25 x 20) μm, with
a 2-layered wall 1.3 μm thick, without micropyle or polar granule,
with (?) residuum; sporocysts spherical or oval, spherical ones 11–
13 μm in diameter, oval ones 14–17 x 11–14 μm.

Sporulation. 2 days at 25 C in 2.5% potassium bichromate solution (Bledsoe, 1976); 2–3 days (Nukerbaeva and Svanbaev, 1973).

Prepatent Period. 6 days in silver fox (Dunlap, 1956); 6–7 days
in dog (Bledsoe, 1976).

Patent Period. 5 days in silver fox (Dunlap, 1956); 7 days in dog
(Bledsoe, 1976).

Cross-Transmission Studies. Nieschulz and Bos (1933) were unable to transmit this species to a young dog of unknown history.
Mantovani (1965) was unable to transmit it to another young dog.
However, Bledsoe (1976, 1976a) transmitted it readily from the
silver fox to the dog (beagle puppies), either directly by means of
sporulated oocysts, or by feeding them mice that had been fed the
oocysts 5–6 months previously. The pups began passing oocysts of
I. vulpina 5 days after having been fed these mice. Bledsoe (1976a)
also transmitted it from dog to dog. Nukerbaeva and Svanbaev
(1974) transmitted this species from *V. vulpes* to *A. lagopus* and
vice versa, but were unable to transmit it from *V. vulpes* to the cat.

Remarks. Although Mantovani (1965) thought the forms he
saw differed in oocyst and sporocyst size from the form that Nieschulz and Bos (1933) had described, it does not appear to us that
the difference was significant. He introduced the subspecific name
vulpina for the farm fox form described by Nieschulz and Bos, and

the subspecific name *aprutina* for the form he saw in feral foxes, but this separation does not appear justified.
See also under *I. canivelocis.*

Isospora pavlodarica Nukerbaeva and Svanbaev, 1973

Type Host. Silver fox *Vulpes vulpes* (synonym, *V. fulva*).
Other Host. Arctic fox *Alopex lagopus.*
Location. Intestine.
Geographic Distribution. USSR (Kazakhstan).
Prevalence. Nukerbaeva and Svanbaev (1973) found this species in 1.6% of 1,199 silver foxes in Kazakhstan.
Oocyst Structure. Short oval or spherical, spherical ones 20–22 μm in diameter, short-oval ones 22–25 x 20–22 μm, with 2-layered wall 1.4 μm thick, without micropyle or residuum, with polar granule; sporocysts ellipsoidal ("oval"), 13 x 7 μm, with residuum, without Stieda body; sporozoites elongate, with one end rounded and the other pointed, with clear globule in broad end.
Sporulation. 3–4 days (Nukerbaeva and Svanbaev, 1973).

Isospora triffitti Nukerbaeva and Svanbaev, 1973

Type Host. Red fox *Vulpes vulpes* (synonym, *V. fulva*).
Other Host. Arctic fox *Alopex lagopus.*
Location. Intestine.
Geographic Distribution. USSR (Kazakhstan).
Prevalence. Nukerbaeva and Svanbaev (1973) found this species in 1.4% of 1,199 silver foxes in Kazakhstan.
Oocyst Structure. Short oval or spherical, spherical ones 11–13 μm in diameter, short-oval ones 11–13 x 10–11 μm, with 2-layered wall 1 μm thick, without micropyle, residuum, or polar granule; sporocysts ellipsoidal ("oval"), 6 x 4 μm, without Stieda body, with residuum; sporozoites elongate, with one end rounded and the other pointed, with clear globule in broad end.

Klossia sp. Golemansky, 1975

(Plate 2, Fig. 11)

Type Host. Red fox *Vulpes vulpes.*
Location. Unknown; oocysts found in large intestine.
Geographic Distribution. Europe (Bulgaria).
Prevalence. Golemansky (1975) found this species in 3 (2%) of 146 red foxes in Bulgaria. One was 1 month old, and 2 were adults.
Oocyst Structure. Ellipsoidal, 30–52 x 25–35 (mean, 38 x 33) μm, with colorless, 2-layered wall about 1.5 μm thick, without micropyle or polar granule, without residuum but with many small bodies that look like drops of fat among oocysts, with 5–16 sporocysts. Sporocysts spherical, 11–12 μm in diameter, without Stieda body but with many residual granules, with 4 sporozoites, most often crossed, 8–10 x 2–3.5 μm. Sporocysts often free from oocysts, already sporulated. Both sporulated and unsporulated oocysts present.
Remarks. Golemansky (1975) thought that this might be a pseudoparasite of the fox.

Sarcocystis citellivulpes Pak, Perminova, and Eshtokina, 1979

Type Definitive Host. Red fox *Vulpes vulpes* (experimental).
Other Definitive Host. Corsac *V. corsac* (experimental).
Intermediate Host. Yellow suslik *Spermophilus* (synonym, *Citellus*) *fulvus.*
Location. Sarcocysts in muscles of suslik, sporocysts in feces of *V. vulpes* and *V. corsac.*
Geographic Distribution. USSR (Kazakhstan).
Prevalence. Pak, Perminova, and Eshtokina (1979) found this species in 4.2% of the yellow susliks that they examined in the Alma-Ata region of Kazakhstan.
Oocyst Structure. Sporocysts ovoid, 10–13 x 7–10 μm.
Sporulation. Occurs in the cells of the definitive host. The feces contain free sporulated sporocysts.
Merogony. According to Pak, Perminova, and Eshtokina (1979),

the sarcocysts in the yellow suslik muscles are 30 x 20–9,000 x
600 μm and the merozoites in them are 9–15 x 3–6 μm.
Prepatent Period. 7–8 days in red fox and corsac.
Patent Period. 7–14 days in red fox and corsac.

Sarcocystis sp. Biocca et al. 1975

Type Definitive Host. Red fox *Vulpes vulpes.*
Other Definitive Hosts. Canadian timber wolf *Canis lupus* and
dog *C. familiaris.* Biocca et al. (1975) infected both of these hosts
and *V. vulpes* but not the cat, lion, ferret, or kestrel *Falco tinnunculus* by feeding them sarcocyst-infected esophageal heart, diaphragm, intercostal, and abdominal muscles from an infected ibex
in Italy.
Intermediate Host. Ibex (steinbok) *Capra ibex.*
Location. Sarcocysts in ibex muscles, sporocysts and oocysts in
fox, wolf, and dog feces.
Geographic Distribution. Europe (Italy).
Prevalence. Unknown. Biocca et al. (1975) found sporocysts in
the feces of an unspecified percentage of 12 red foxes in the Gran
Paradiso National Park, Italy.
Oocyst Structure. Sporocysts 13–15 x 8–10 μm, with residuum,
without Stieda body.
Sporulation. Occurs in the cells of the definitive host. The feces
contain a few sporulated oocysts and numerous sporulated sporocysts.
Merogony. Not described. Sections of sarcocysts in muscles with
a thick wall containing cytophaneres illustrated.
Prepatent Period. 11 days in the fox, 12 days in the wolf, 20
days in the dog (Biocca et al., 1975).
Patent Period. 62 days in the fox, 67 days in the wolf, 66 days
in the dog (Biocca et al., 1975).

Host Genus *Fennecus*

Isospora fennechi Prasad, 1961

(Plate 2, Fig. 10)

Type Host. Fennec fox *Fennecus zerda.*
Location. Unknown; oocysts found in feces.
Geographic Distribution. Europe (London zoo).
Oocyst Structure. Oocysts subspherical, ovoid, or cylindrical, 24–31 x 15–20 (mean, 27.5 x 14) μm, with thin wall, without micropyle, residuum, or polar granule. Sporocysts ovoid, 15–17 x 10–12 μm, with Stieda body, with residuum. Sporozoites club-shaped, about 14 x 5 μm.
Sporulation. Freshly passed oocysts already contained 2 sporoblasts. Remaining time, 24 hours.
Cross-Transmission Studies. Prasad (1961) was unable to infect 2 puppies with this species.

Host Genus *Nyctereutes*

Sarcocystis sp. Britov, 1970

Definitive Host. Unknown.
Intermediate Host. Raccoon dog *Nyctereutes procyonoides.*
Location. Striated muscles.
Geographic Distribution. USSR (Primorye).
Merogony. Only sarcocysts known.

Host Family URSIDAE

Host Genus *Ursus*

Eimeria ursi Yakimoff and Matchulski, 1935

(Plate 15, Figs. 79, 80)

Type Host. Brown bear *Ursus arctos.*
Location. Unknown; oocysts found in feces.
Geographic Distribution. USSR (Leningrad zoo).
Oocyst Structure. Oocysts spherical to ovoid, ovoid oocysts 13–15 x 11–13 (mean, 14 x 11.5) μm, with hyaline wall, with micropyle and polar granule, without residuum. Sporocysts 6 x 4 μm, without Stieda body or residuum.

Sporulation. 3 days at 24 C in 2% potassium bichromate solution.
Cross-Transmission Studies. Yakimoff and Matchulski (1935)
were unable to infect a kitten with this species.

Isospora fonsecai Yakimov and Machul'skii, 1940

(Plate 4, Fig. 21)

Synonym. Isospora lacazei Labbé (1899) of Yakimoff and
Matchulski (1935).
Type Host. Red bear *Ursus arctos isabellinus* (synonym, *U.
pamirensis*).
Location. Unknown; oocysts found in feces.
Geographic Distribution. USSR (Tashkent, Leningrad, and
Kharkhov zoos).
Oocyst Structure. Oocysts spherical, 18–26 (mean, 23) μm in
diameter or subspherical, 22–32 x 20–30 (mean, 25 x 23) μm, with
pinkish wall about 1 μm thick, without micropyle or residuum,
with polar granule. Sporocysts piriform, 14–18 x 8–12 μm, with
Stieda body and residuum.
Sporulation. 4 days at 24 C in 2% potassium bichromate solution.
Cross-Transmission Studies. Yakimoff, Iwanoff-Gobzem, and
Matschoulsky (1936) were unable to infect a kitten with this
species.
Remarks. Perhaps this is the same species that Yakimoff and
Matchulski (1935) called *"Isospora lacazei* Labbe" because they
were struck with its resemblance to the oocysts that Yakimoff and
Gusev (1936a) had found in sparrows in the USSR. This species
was found in the feces of the brown bear *U. arctos* in the Leningrad zoo. At that time it was not realized that the brown bear was
the same species as the red bear, which Yakimoff and Machul'skii
(1940) later called *U. pamirensis.* The oocysts of *"I. lacazei"* from
U. arctos were spherical, 19–32 (mean, 25) μm in diameter, or
ovoid, 23–34 x 21–29 (mean, 26 x 24) μm, with a wall 1.5 μm
thick, without micropyle, but with residuum composed of a few
granules and polar granule, with sporocysts 15–19 x 11–13 μm,
with a Stieda body, substiedal body, and residuum, with semilunar
sporozoites.

Host Genus *Euarctos*

Eimeria albertensis Hair and Mahrt, 1970

(Plate 6, Fig. 36)

Type Host. Black bear *Euarctos americanus.*
Location. Unknown; oocysts found in feces.
Geographic Distribution. North America (Alberta).
Prevalence. Hair and Mahrt (1970) found this species in 8% of
52 black bears in Alberta.
Oocyst Structure. Oocysts ellipsoidal, 36–44 x 19–23 (mean,
41.5 x 22) μm, with smooth, brownish, 2-layered wall, outer layer
about 0.5 μm thick and thickened around micropyle, inner about
1 μm thick, with micropyle about 5 μm wide, usually with residuum, without polar granule. Sporocysts ovoid to elongate ovoid,
11–19 x 7–11 (mean, 15 x 8) μm, with Stieda body and residuum.
Sporozoites vermiform, lying lengthwise head to tail in sporocysts,
with prominent clear globule.

Eimeria borealis Hair and Mahrt, 1970

(Plate 6, Fig. 31)

Type Host. Black bear *Euarctos americanus.*
Location. Unknown; oocysts found in feces.
Geographic Distribution. North America (Alberta).
Prevalence. Hair and Mahrt (1970) found this species in 6% of
52 black bears in Alberta.
Oocyst Structure. Oocysts ellipsoidal to slightly concave on one
side, 29–33 x 15–16 (mean, 30 x 15) μm, with smooth, colorless,
2-layered wall, outer layer about 0.8 μm thick, inner about 0.2 μm
thick, wall thin and slightly flattened at anterior end, with indistinct
micropyle about 3 μm wide and residuum, without polar granule.
Sporocysts ovoid, 7–12 x 4–7 (mean, 10 x 6) μm, with Stieda body
and residuum. Sporozoites vermiform, lying lengthwise head to tail
in sporocysts, with clear globule at one end.

Sarcocystis sp. Crum, Nettles, and Davidson, 1978

This species was found in the striated and/or heart muscles of 11% of 53 black bears in the southeastern United States by Crum, Nettles, and Davidson (1978).

Host Family PROCYONIDAE

Host Genus *Procyon*

Eimeria nuttalli Yakimoff and Matikaschwili, 1933

(Plate 5, Fig. 28; Plate 16, Figs. 84, 85)

Host Type. Raccoon *Procyon lotor.*
Location. Unknown; oocysts found in feces.
Geographic Distribution. USSR (Leningrad—raccoons from North America), Europe (England), North America (Alabama, Georgia, Iowa, Michigan).
Prevalence. Inabnit, Chobotar, and Ernst (1972) found this species in 21% of 48 raccoons from Alabama, Georgia, and Michigan.
Oocyst Structure. Ellipsoidal, 16–23 x 13–16 (mean, 19.5 x 14) μm, with 2-layered wall, outer layer 1.3–1.7 μm thick, inner thinner, without micropyle or residuum, at times with polar granule. Sporocysts ovoid or ellipsoidal, 7–9 x 5–7 μm, without Stieda body, with residuum. Sporozoites comma-shaped.

The above is from Yakimoff and Matikaschwili (1933). According to Morgan and Waller (1940) the oocysts in a raccoon in Iowa were ovoid, 20 x 14 μm, with a "double contoured" wall, without a micropyle, mostly with a polar granule, and the sporocysts were ovoid. According to Inabnit, Chobotar, and Ernst (1972), who reported *E. nuttalli* in raccoons from Alabama, Georgia, and Michigan, the oocysts were ovoid, 12–21 x 11–15 (mean, 17 x 14) μm, with a 1-layered, smooth, yellowish-brown wall about 0.7 μm thick, without micropyle or residuum, with polar granule, and the sporocysts were ovoid, 9–13 x 5–11 (mean, 12 x 7) μm, with a small dark Stieda body and a residuum.

Eimeria procyonis Inabnit, Chobotar, and Ernst, 1972

(Plate 7, Fig. 38)

Type Definitive Host. Raccoon *Procyon lotor.*
Location. Unknown; oocysts found in feces.
Geographic Distribution. North America (Michigan, Alabama, and Georgia).
Prevalence. Inabnit, Chobotar, and Ernst (1972) found this species in 67% of 48 raccoons from Michigan, Alabama, and Georgia.
Oocyst Structure. Ellipsoidal to ovoid, 16–29 x 13–24 (mean, 23 x 18) μm with 2-layered wall, outer layer rough, pitted, yellowish-brown, radially striated, 1.5 μm thick, except at micropylar end, where it is 1 μm thick, inner layer colorless to yellow, 0.5 μm thick, with micropyle and polar granule, without residuum. Sporocysts ovoid, 11–15 x 7–10 (mean, 12 x 9) μm, with Stieda body, substiedal body, and residuum. Sporozoites with very large clear globules.

Isospora chobotari n. sp.

(Plate 7, Fig. 37)

Synonym. Isospora sp. Inabnit, Chobotar, and Ernst, 1972.
Type Definitive Host. Raccoon *Procyon lotor.*
Location. Unknown; oocysts found in feces.
Geographic Distribution. North America (Michigan).
Prevalence. Inabnit, Chobotar, and Ernst (1972) found this species in 1 of 43 raccoons from Michigan.
Oocyst Structure. Ellipsoidal, 16–19 x 12–16 (mean, 17 x 14) μm, with smooth, brown, 1-layered wall 1 μm thick, without micropyle, residuum, or polar granule. Sporocysts ellipsoidal, 9–12 x 8–10 (mean, 11 x 9) μm, without Stieda body, with residuum. Sporozoites elongate, with one end blunt and the other pointed, without clear globules.
Remarks. While Inabnit, Chobotar, and Ernst (1972) did not name this species, they indicated that it differed from all other

species of *Isospora* occurring in carnivores. We are naming it in honor of Dr. Chobotar.

Sarcocystis leporum Crawley, 1914

See under *Felis catus.*

Sarcocystis sp. Seneviratna, Edward, and DeGiusti, 1975

Seneviratna, Edward, and DeGiusti (1975) found sarcocysts in the muscles of 3 of 6 raccoons *Procyon lotor* in the Detroit, Michigan, area. They thought that this was the first record of *Sarcocystis* in the raccoon.

Host Genus *Nasua*

Eimeria nasuae Carini and Grechi, 1938

(Plate 6, Fig. 32)

Type Host. Coatimundi *Nasua nasua.*
Location. Unknown; oocysts found in feces.
Geographic Distribution. South America (Brazil).
Oocyst Structure. Subspherical, 17–19 x 15–17 μm, with rather rough, radially striated, double-contoured wall, without micropyle or residuum. Sporocysts ovoid, about 10 x 8 μm, without Stieda body, with residuum.
Sporulation. 10–11 days at room temperature in 1% chromic acid solution.
Immunity. Carini and Grechi (1938) thought that their failure to infect an older coatimundi with oocysts from the original one was due to immunity.

Host Genus *Potos*

Eimeria poti Lainson, 1968

(Plate 8, Fig. 45)

Type Host. Kinkajou *Potos flavus.*
Location. Unknown, presumably intestine.
Geographic Distribution. Central America (British Honduras).
Prevalence. Lainson (1968) found this species in 1 out of 5 *P. flavus* in British Honduras.
Oocyst Structure. Ellipsoidal to cylindroid, 26–31 x 17–20 (mean, 30 x 18) μm, with colorless, smooth, 2-layered wall about 0.8 μm thick, without micropyle, residuum, or polar granule. Sporocysts ellipsoidal, 14–15 x 8–9 (mean, 15 x 9) μm, with a pointed Stieda body at what was described as the narrow end, with a residuum. Sporozoites elongate, recurved, lying lengthwise head to tail in sporocysts; presence of clear globule uncertain.
Sporulation. 6 days at 26–28 C in 2% potassium bichromate solution.

Host Family MUSTELIDAE

Host Genus *Mustela*

Eimeria mustelae Iwanoff-Gobzem, 1934

(Plate 10, Fig. 57)

Type Host. Snow weasel *Mustela nivalis.*
Other Host. Mink *M. vison.*
Location. Feces of snow weasel; duodenum and ileum lumina of mink.
Geographic Distribution. USSR (Kazakhstan), North America (Illinois).
Oocyst Structure. According to Iwanoff-Gobzem (1934) and Ivanova-Gobzem (1935), the oocysts in the snow weasel were spherical, 18–26 (mean, 20) μm in diameter, or ellipsoidal, 18–26 x 14–24 (mean, 22 x 19) μm, with a fairly thick, double-contoured wall, without a micropyle or residuum, with a polar granule. Sporocyst residuum absent. According to Levine (1948), the oocysts in the mink were subspherical, 13–18 x 12–15 (mean, 16 x 14) μm, with a 2-layered wall, outer layer colorless and inner pale salmon pink, without micropyle or residuum, with polar granule. Sporocysts roughly ovoid but with a constriction at one end, 8 x 5 μm,

with Stieda body and residuum. Sporozoites 7 x 3 µm, broad at one end and tapering toward the other, lying lengthwise head to tail in sporocysts, apparently without clear globule.

Sporulation. 3 days at room temperature in 2.5% potassium bichromate solution.

Remarks. Levine (1948) found this species in a single mink from a fur farm in Illinois.

Eimeria furonis Hoare, 1927

(Plate 6, Figs. 33, 34, 35)

Type Host. Ferret *Mustela putorius* var. *furo.* Hoare (1927) emphasized that this was the name of the ferret, which is a domesticated variety of the polecat *M. (P.) putorius,* and that both occur only in Europe.

Other Host. Mink *M. vison.*

Location. Above the host cell nuclei of the epithelial cells of the small intestine and rectum, mostly in the apices of the villi and not in the crypts of Lieberkuehn.

Geographic Distribution. Europe (England), USSR (Kazakhstan).

Prevalence. Nukerbaeva and Svanbaev (1973) reported this species from 1.7% of 1,027 mink in Kazakhstan.

Oocyst Structure. Spherical to subspherical, 11–14 x 10–13 (mean, 13 x 12) µm, with a 2-layered wall, outer layer colorless and thin, inner yellowish and thick, without micropyle, residuum, or apparently polar granule. Sporocysts irregularly spindle-shaped, one end being slightly constricted and blunted, 8–9 x 4 µm, apparently with Stieda body, with residuum. Sporozoites vermiform with one end narrower than the other, lying lengthwise head to tail in sporocysts, sometimes with clear globule at broad end. The oocysts reported by Nukerbaeva and Svanbaev (1973) from mink in Kazakhstan were short oval or spherical, the spherical ones 11 µm in diameter, the short-oval ones 12–14 x 10–13 (mean, 13 x 11) µm, with a 2-layered wall 1 µm thick, without micropyle or polar granule; sporocysts 6 x 4 µm.

Sporulation. 4–6 days at room temperature in 0.5% chromic acid; 5–6 days (Nukerbaeva and Svanbaev, 1973).

Merogony. The earliest stages are spherical and 3–4 μm in dia- meter, with a relatively large nucleus containing a deeply staining karyosome. Hoare (1927) saw 2 types of merogony. In the first, stumpy, sausage-shaped merozoites about 3–4 x 2 μm were budded off, leaving a cytoplasmic residuum; he illustrated about 14 mero- zoites in a meront section. In the second, the merozoites were elongate and curved, about 6 x 1 μm, with one end rounded and the other drawn out, and with a compact nucleus near the rounded end; in this type the merozoites appeared to lie within the meront rather than being budded off from it. Hoare (1927) thought that these merozoites gave rise to gamonts.

Gametogony. According to Hoare (1927) the macrogametes are spherical, about 8 μm in diameter, and contain darkly staining globules. The microgamonts are about the same size.

Prepatent Period. 5 days.

Pathogenicity. According to Hoare (1927) heavily infected ferrets had no signs of disease. The only effect was enlargement and irregular arrangement of the epithelial cells in the most heavily infected parts of the mucosa and denudation of some areas due to shedding of infected epithelium.

Remarks. This species was found in a laboratory ferret.

Eimeria ictidea Hoare, 1927

(Plate 10, Fig. 56)

Type Host. Ferret *Mustela putorius* var. *furo;* polecat *M. putor- ius.*

Other Host. Steppe weasel *M. eversmanni* (?).

Location. Above the host cell nuclei in the epithelial cells of the free part of the villi, especially the tip, in the small intestine.

Geographic Distribution. Europe (England), USSR (Byelorussia, ?Kazakhstan).

Oocyst Structure. The following description is by Hoare (1927) of the oocysts in the ferret. Oocysts elongate ovoid or ellipsoidal, 13–27 x 13–21 (mean, 24 x 17.5) μm, with 2-layered wall, outer layer colorless, thin, inner layer slightly yellowish, thick, with small micropyle sometimes visible, without residuum, with polar granule apparently disappearing before sporulation is complete. Sporocysts irregularly ovoid, with one end broad and rounded and the other slightly constricted, about 11.5 x 6.5 μm, apparently with Stieda body, with residuum. Sporozoites vermiform, narrowed at one end, lying lengthwise head to tail in sporocysts, with clear globule at broad end. Litvenkova (1969) did not describe the *E. ictidea* she found in 1 of 3 polecats in Byelorussia. Svanbaev (1956) gave the following description of the oocysts he found in *M. eversmanni* in Kazakhstan; the sporocysts and sporozoites differ from those described by Hoare, and Svanbaev may have been dealing with a different species. Oocysts ellipsoidal to short ellipsoidal, 25–28 x 20–21 (mean, 26 x 20) μm, with smooth, colorless, double-contoured wall 1.1–1.3 μm thick, without micropyle, residuum, or polar granule. Sporocysts spherical or short ellipsoidal, 8–9 μm in diameter (mean, 9 x 8.5 μm), with residuum. Sporozoites, 6 x 4 μm.

Sporulation. 7 days (Hoare, 1927) or 40–44 hours at 20–25 C in potassium bichromate solution (Svanbaev, 1956).

Merogony. Hoare (1927) described the endogenous stages. The youngest stages he saw were merozoites free in the lumen of the gut near groups of fully segmented intracellular meronts from which they had arisen. They were elongated and vermicular, about 11 x 1 μm, with one end rounded and the other pointed. The nucleus was compact and near the rounded end. The merozoites rounded up after penetrating a host cell, grew, and became 3–4 μm in diameter. Hoare illustrated 14 in a meront, but there were probably more. They apparently turned into gamonts.

Gametogony. The mature macrogametes and microgamonts were about 20 x 7 μm.

Prepatent Period. 4 days.

Pathogenicity. Hoare (1927) saw no signs of disease in heavily infected ferrets. An annular constriction extending into the villar core separated the infected from the healthy part of the affected villi. The blood vessels of the lamina propria of the infected part of the villus were dilated and engorged with erthrocytes, and there was some extravasation into the surrounding tissue and also some necrosis, but infected ferrets remained clinically healthy.

Remarks. Hoare (1927) found this species in a laboratory ferret.

Eimeria vison Kingscote, 1935

(Plate 2, Fig. 12)

Synonym. Eimera mustelae Kingscote, 1934.
Type Host. Mink *Mustela vison.*
Other Host. Ferret *M. putorius* var. *furo.*
Location. Small intestine, perhaps large intestine.
Geographic Distribution. North America (Ontario, Illinois, Wisconsin), Europe (England), USSR (Kazakhstan).

Prevalence. Foreyt and Todd (1976) found this species in mink on 62% of 29 mink ranches in Wisconsin. McTaggart (1960) found it in 3 of 200 mink from 42 farms in Britain. Nukerbaeva and Svanbaev (1973) found this species in 2.2% of 1,027 mink in Kazakhstan.

Oocyst Structure. According to Kingscote (1934) the oocysts are ovoid, 17–22 x 9–18 (mean, 20 x 15) μm, with a 2-layered wall 0.75 μm thick, outer layer colorless, inner layer yellowish brown and relatively thick, without micropyle, with residuum in some oocysts. Sporocysts ovoid at first and then piriform, 10 x 5.5 μm, without Stieda body, with residuum. Sporozoites slightly curved, club-shaped, tapered at one end, 9 x 2.5 μm, lying head to tail in sporocysts. Levine (1948) described the oocysts as ellipsoidal, 20–26 x 13–17 (mean, 23 x 15) μm, with 2-layered wall, outer layer colorless, inner pale salmon pink, without micropyle or residuum, with polar granule. Sporocysts ellipsoidal, 11 x 8 μm, without Stieda body, with residuum. Sporozoites comma-shaped, 9 x 3 μm, lying head to tail in sporocysts, with clear globule.

According to McTaggart (1960), the oocysts were usually ellipsoidal, sometimes ovoid and occasionally subspherical, 16–26 x 12–17 (mean, 22 x 15) μm, with a 2-layered wall, outer layer colorless, inner pale straw-colored, without micropyle or residuum, with a sporocyst residuum. The oocysts described by Nukerbaeva and Svanbaev (1973) in mink in Kazakhstan were short oval or ellipsoidal, the short-oval ones 20–22 x 14–15 μm, the ellipsoidal ones 28 x 15 μm, with a 2-layered wall 1.5 μm thick, without micropyle; sporocysts 7–10 x 6 μm, with residuum; sporozoites 6 x 3 μm.

Sporulation. 7 days at room temperature in 2.5% potassium bichromate solution (Kingscote, 1934); 3 days under the same conditions (Levine, 1948); about 5 days at room temperature (McTaggart, 1960); 2-3 days (Nukerbaeva and Svanbaev, 1973).

Prepatent Period. 5 days (Kingscote, 1934).

Patent Period. 4 days (Kingscote, 1934).

Pathogenicity. According to Kingscote (1934), infected mink from a fur farm died of enteric coccidiosis due to this species.

Cross-Transmission Studies. Kingscote (1934) infected 2 ferrets with this species but was unable to infect 2 kittens, a rabbit, and a guinea pig. Nukerbaeva and Svanbaev (1974) were unable to transmit this species from the mink to *Vulpes vulpes, Alopex lagopus,* or the dog.

Eimeria hiepei Gräfner, Graubmann, and Dobbriner, 1967

(Plate 7, Fig. 44; Plate 8, Fig. 49)

Type Host. Mink *Mustela* (synonym, *Lutreola*) *vison.*

Location. Bile duct epithelium.

Geographic Distribution. Europe (East Germany), North America (Wisconsin).

Oocyst Structure. Oocysts spherical, 13-17 μm in diameter, with smooth, colorless, 2-layered wall, without micropyle, residuum, or polar granule. Sporocysts about 6 x 4 μm, apparently without Stieda body, without residuum. Sporozoites banana-shaped, with one end pointed and the other broadly rounded.

Sporulation. 2 days at 24 C in 2.5% potassium bichromate solution.

Merogony. Gräfner, Graubmann, and Dobbriner (1967) said that there were type A and type B meronts in the bile duct epithelial cells like those of *E. stiedai* of the rabbit. The former had 8 merozoites and the latter 16. They thought that type A meronts formed first, but they were not sure. At any rate, these meronts reached a maximum of about 8 μm in diameter, while the spherical meronts of type B were 12-15 μm in diameter. Both types of merozoite were banana-shaped, about 8 x 2 μm.

Gametogony. The macrogametes were spherical to ovoid, with a diameter of 8–10 μm. The microgamonts were spherical, with a mean diameter of about 8 μm.

Pathogenicity. Gräfner, Graubmann, and Dobbriner (1967) said that the livers of affected mink contained yellowish, irregular nodules or hollow structures from the size of a peppercorn to that of a pea. The bile ducts in the affected areas of the liver contained detritus, leukocytes, gamonts, oocysts, meronts, and many eosinophils. The epithelial cells of the bile duct were often swollen, and there was often proliferation of the mucosa of the main bile ducts. The mink described by Davis, Chow, and Gorham (1953) on a fur farm in Wisconsin died with lesions reminiscent of a neoplasm of the liver. The bile ducts were thickened and resembled coral.

Remarks. Pellérdy (1974) was not sure whether the host was the European mink *M. l. lutreola* (which Walker et al., 1975, call *M. lutreola*) or the American mink *M. l. vison* (which Walker et al., 1975, call *M. vison*).

This is undoubtedly the species that Davis, Chow, and Gorham (1953) found in a fur farm mink in Wisconsin.

Isospora laidlawi Hoare, 1927

(Plate 9, Figs. 50, 54; Plate 10, Fig. 59)

Type Host. Ferret *Mustela putorius* var. *furo.*
Other Host. Mink *M. vison.*
Location. Unknown; oocysts found in feces or intestinal contents.
Geographic Distribution. Europe (England), USSR (Kazakhstan), North America (Illinois, Wisconsin).
Prevalence. McTaggart (1960) found this species in 5 mink out of 200 from 42 mink farms in Britain. Foreyt and Todd (1976) found it in 90% of 29 mink ranches in Wisconsin. Nukerbaeva and Svanbaev (1973) found it in 1.9% of 1,027 mink in Kazakhstan.
Oocyst Structure. The oocysts from the ferret were described by Hoare (1927). Ovoid, 32–37 x 27–30 (mean, 34 x 29) μm, with a 2-layered wall, the outer layer colorless and thin, the inner layer

yellowish and thick, without micropyle, residuum, or apparently polar granule. Sporocysts ellipsoidal, 21 x 14 μm, without Stieda body, with residuum. Sporozoites sausage-shaped, with one end more pointed than the other, lying lengthwise and pointing in the same direction in the sporocysts, with a clear globule at the small end. Levine (1948) described the oocysts from the mink as ellipsoidal, 32–36 x 26–27 (mean, 34 x 26.5) μm, with 2-layered wall, the outer layer colorless, the inner layer pale yellow, without micropyle, residuum, or polar granule. Sporocysts ellipsoidal, without Stieda body, with residuum. Sporozoites sausage-shaped. McTaggart (1960) described the oocysts from the mink as mostly ovoid, sometimes ellipsoidal and occasionally subspherical, 30–39 x 25–34 (mean, 35 x 29) μm, with 2-layered wall, outer layer colorless, inner layer almost colorless to very pale yellow, without micropyle or residuum. Sporocysts 19–28 x 14–20 (mean, 23 x 17) μm, with residuum. The oocysts seen by Nukerbaeva and Svanbaev (1973) in mink in Kazakhstan were short oval, 31–37 x 25–31 (mean, 34 x 28) μm, with 2-layered wall, without micropyle or polar granule; sporocysts spherical 14–15 μm in diameter, with residuum.

Sporulation. 4 days at room temperature in 0.5% chromic acid solution (Hoare, 1927); 3 days at room temperature (McTaggart, 1960); 2 days (Nukerbaeva and Svanbaev, 1973).

Cross-Transmission Studies. Nukerbaeva and Svanbaev (1974) were unable to transmit this species from the mink to *Vulpes vulpes, Alopex lagopus,* or the dog.

Isospora eversmanni Svanbaev, 1956

(Plate 15, Fig. 82)

Type Host. Eversmann's polecat *Mustela eversmanni.*
Other Host. Mink *M. vison.*
Location. Unknown; oocysts found in feces.
Geographic Distribution. USSR (Kazakhstan).
Prevalence. Nukerbaeva and Svanbaev (1973) found this species in 0.2% of 1,027 mink in Kazakhstan.

Oocyst Structure. Oocysts spherical, 20 μm in mean diameter, with smooth, colorless, double-contoured wall about 1 μm thick, without micropyle, residuum, or polar granule. Sporocysts spherical or ellipsoidal, 10 μm in diameter, apparently without Stieda body, without residuum. Sporozoites ovoid, averaging 5 x 3 μm. The oocysts seen in mink by Nukerbaeva and Svanbaev (1973) were spherical, 17–22 μm in diameter, with a 2-layered wall 1.4 μm thick, without micropyle or polar granule, with residuum; sporocysts 11 x 8 μm.

Sporulation. 43–45 hours at 20–25 C in potassium bichromate solution.

Isospora pavlovskyi Svanbaev, 1956

(Plate 16, Fig. 87)

Type Host. Evermann's polecat *Mustela eversmanni.*
Location. Unknown; oocysts found in feces.
Geographic Distribution. USSR (Kazakhstan).
Oocyst Structure. Oocysts ovoid, 37–40 x 29–31 (mean, 38.5 x 30) μm, with smooth, colorless, double-contoured wall 1.6–1.8 μm thick, without micropyle or polar granule, with or without residuum. Sporocysts spherical or ellipsoidal, 19 x 15–17 (mean, 19 x 16) μm, apparently without Stieda body, with residuum. Sporozoites ellipsoidal, 6–7 x 5–6 (mean, 7 x 5) μm.
Sporulation. 38–45 hours at 20–25 C in potassium bichromate solution.

Sarcocystis putorii (Railliet and Lucet, 1891)
Tadros and Laarman, 1978

(Plate 10, Fig. 58)

Synonyms. Endorimospora putorii Tadros and Laarman, 1976; *Coccidium bigeminum* var. *putorii* Railliet and Lucet, 1891 *nomen nudum; Isospora putorii* (Railliet and Lucet, 1891) Becker, 1934 of Becker (1934) and *auctores (nomen nudum).*

Type Definitive Host. Common European weasel *Mustela nivalis.*
Other Definitive Hosts. Ferret *M. putorius* var. *furo,* stoat *M. erminea,* and (?) mink *M. vison.*
Intermediate Host. Common European vole *Microtus arvalis.*
Other Intermediate Host. Short-tailed vole *M. agrestis.*
Location. Gamonts, gametes, zygotes, oocysts, and sporocysts in lamina propria of small intestine of mustelids. Initial meronts in liver and other visceral organs of voles. Sarcocysts in muscles of voles.
Geographic Distribution. Europe (Netherlands, probably Switzerland and elsewhere), probably in mink in North America (Illinois).
Oocyst Structure. Oocysts described only by Levine (1948) in scrapings from duodenal mucosa of a mink. Sporulation occurs in the intestine. Oocysts 11–14 x 7–11 (mean, 12 x 9) μm, with smooth, 1-layered wall so thin as to be almost invisible except under oil immersion objective, stretched by sporocysts within it, without micropyle, residuum, or polar granule. Sporocysts ellipsoidal, about 11 x 7 μm, with 2-layered wall, without Stieda body, with large residuum. The sporocysts seen by Tadros and Laarman (1976) in *M. nivalis* were 11–13 x 8–10 μm, with a granular residuum.
Sporulation. Occurs in intestine of definitive host. Sporulated sporocysts passed in feces.
Merogony. According to Tadros and Laarman (1976) the first-generation meronts develop in the liver and other visceral organs. The last-generation meronts (sarcocysts) are up to several centimeters long, are compartmented, and contain both metrocytes and bradyzoites. The sarcocyst wall has sparse villuslike projections up to 3 μm long, with numerous fibrils visible with the electron microscope.
Prepatent Period. 7–13 days (Tadros and Laarman, 1976).
Patent Period. 2 weeks to 3 months (Tadros and Laarman, 1976).
Pathogenicity. Not pathogenic for mustelids. Heavy infections of voles may cause loss of appetite, lethargy, excessive thirst, muscular weakness, or even death (Tadros and Laarman, 1976).
Cross-Transmission Studies. Tadros and Laarman (1976) found sporocysts of this species in the feces of *Mustela nivalis,* fed them to the vole *Microtus arvalis,* produced sarcocysts in it, and then

infected not only the weasel but also the stoat and ferret. They also infected *Microtus agrestis,* but not *Clethrionomys glareolus, Mesocricetus auratus, Mus musculus,* or *Rattus norvegicus.*

Remarks. This is probably the same organism that Galli-Valerio (1935) called *"Isospora bigemina* var. *putori"* in a *"Mustela herminea"* in Switzerland, and it may well be the *"Isospora bigemina"* that Levine (1948) found in the intestine of a fur farm mink *M. vison* in Illinois.

Sarcocystis sp. Tadros and Laarman, 1979

Tadros and Laarman (1979) found sarcocysts of this species in the muscles of the common European weasel *Mustela nivalis* in the Netherlands. The sarcocysts were up to several mm long by 0.15 mm in diameter, with a smooth wall without cytophaneres, compartmented, with metrocytes 3.5 μm in diameter and bradyzoites averaging 9 x 2.5 μm (much like the sarcocysts of *S. sebeki* in *Apodemus sylvaticus*). The same weasel had sporulated oocysts of *S. putorii* in its feces. They obtained a few oocysts for a short time in a tawny owl *Strix aluco* to which they had fed the weasel sarcocysts, but they thought that this was an abnormal host and that another genus of strigid bird might be a more favorable definitive host. They came to no conclusion as to what species the weasel sarcocysts belonged.

Host Genus *Martes*

Eimeria sibirica Yakimoff and Terwinsky, 1931

Type Host. Sable *Martes zibellina.*

Location. Unknown; oocysts found in feces or intestinal contents.

Geographic Distribution. USSR (Siberia, Baku [Azerbaidzhan] zoo).

Oocyst Structure. Oocysts ellipsoidal, 20–28 x 16–21 μm, with double-contoured wall, without micropyle, polar granule, or residuum. No further structural information available; sporulation did not proceed beyond the 4-sporoblast stage.

Eimeria sp. Yakimoff and Gousseff, 1934

Type Host. Sable or marten *Martes martes.*
Location. Unknown; oocysts found in feces.
Geographic Distribution. USSR (Azerbaidzhan).
Oocyst Structure. Ovoid, 20–31 x 16–20 (mean, 22 x 18) μm, without micropyle, residuum, or polar granule. Sporoblasts said to be spherical, 7 μm in diameter but illustrated as ellipsoidal, presumably without Stieda body, with residuum. Sporozoites 13 x 6 μm. No other structural information given.
Sporulation. More than 2 days.

Isospora mustelae Galli-Valerio, 1932

Host. Marten *Martes martes.*
Nomen nudum (see Pellérdy, 1974).

Host Genus *Eira*

Eimeria irara Carini and da Fonseca, 1938

(Plate 3, Fig. 17)

Type Host. Tayra *Eira* (synonyms, *Tayra, Galictis, Galera*) *barbara.*
Location. Unknown; oocysts found in feces.
Geographic Distribution. South America (Brazil).
Oocyst Structure. Oocysts ovoid, 21–25 x 18–20 μm, with smooth, colorless wall, without micropyle, residuum, or polar granule. Sporocysts ellipsoidal, 10–12 x 6.5 μm, with Stieda body and residuum. Sporozoites elongate, with one end broader than the other.
Sporulation. 2 days in potassium bichromate solution.

Host Genus *Poecilictis*

Isospora africana Prasad, 1961

(Plate 4, Fig. 22)

Type Host. North African weasel *Poecilictis libyca alexandrae.*
Location. Unknown; oocysts found in feces.
Geographic Distribution. Africa (Egypt).
Oocyst Structure. Spherical, 25–27 (mean, 26) μm in diameter, with smooth wall composed of 2 pale yellow membranes, without micropyle, residuum, or polar granule. Sporocysts ovoid, 15–17 x 10–12 μm, without Stieda body, with prominent residuum. Sporozoites elongate, about 13.5 x 3 μm, with one end apparently narrower than the other, with a clear globule at the broad end.
Sporulation. 36 hours at 23 C in 2.5% potassium bichromate solution.
Cross-Transmission Studies. Prasad (1961) was unable to infect a kitten with this species.

Isospora hoogstraali Prasad, 1961

(Plate 7, Figs. 42, 43)

Type Host. North African weasel *Poecilictis libyca alexandrae.*
Location. Unknown; oocysts found in feces.
Geographic Distribution. Africa (Egypt).
Oocyst Structure. Most oocysts ellipsoidal, 37–41 x 32–34 (mean, 38 x 33) μm, with 2-layered wall, outer layer smooth, colorless, relatively thick, inner layer colorless, relatively thin, with button-like micropyle, without residuum, with polar granule in some. Sporocysts ovoid, 19–21 x 13–15 μm, with 2-layered wall, without Stieda body, with residuum. Sporozoites club-shaped, 18–19 x 4–6 μm, with a large clear globule at the broad end.
Sporulation. 1 day at 23 C in 2.5% potassium bichromate solution.
Cross-Transmission Studies. Prasad (1961) claimed to have infected a kitten with this species.

Isospora zorillae Prasad, 1961 emend. Pellérdy, 1963

(Plate 7, Figs. 40, 41)

Synonym. Isospora bigemina var. *zorillae* Prasad, 1961.
Type Host. North African weasel *Poecilictis libyca alexandrae,*
Location. Unknown, oocysts found in feces.
Geographic Distribution. Africa (Egypt).
Oocyst Structure. Most oocysts ovoid, 10–14 x 8–12 (mean, 12
x 10) μm, with thin, 1-layered wall, without micropyle or apparently residuum or polar granule. Sporocysts ovoid, 6–8 x 5 μm,
with 2-layered wall much thicker than oocyst wall, without Stieda
body, with prominent residuum. Sporozoites sickle-shaped, with
one end broadly rounded and the other ending in a blunt point,
6 x 3 μm, with a clear globule at the broad end.
Cross-Transmission Studies. Prasad (1961) was unable to infect
a kitten with this species.

Host Genus *Mellivora*

Sarcocystis sp. Viljoen, 1921

Definitive Host. Unknown.
Intermediate Host. Honey badger or ratel *Mellivora capensis.*
Location. Striated muscles.
Geographic Distribution. Africa (South Africa).
Remarks. The Viljoen (1921) paper was not seen; the above is
from Nietz (1965).

Host Genus *Meles*

Eimeria melis Kotlan and Pospesch, 1933

Type Host. Old World badger *Meles meles.*
Location. Unknown; oocysts found in feces.
Geographic Distribution. Europe (Hungary—Budapest zoo).
Oocyst Structure. Generally ovoid, less frequently subspherical
or ellipsoidal, 17–24 x 13–17 μm, with smooth, colorless, moder-

ately thick wall, without micropyle, with residuum at first. Sporocysts more or less fusiform, with small residuum.

Sporulation. 2–4 days.

Prepatent Period. 4 days.

Cross-Transmission Studies. Kotlan and Pospesch (1933) could
not infect young cats with this species.

Remarks. Pospesch was Pellérdy's name before he changed it.

Isospora melis Pellérdy, 1955

(Plate 13, Fig. 72)

Synonyms. Lucetina sp. Kotlan and Pospesch, 1933; *Isospora
melis* (Kotlan and Pospesch, 1933) Pellérdy, 1955 of Pellérdy
(1965).

Type Host. Old World badger *Meles meles.*

Location. Unknown; oocysts found in feces.

Geographic Distribution. Europe (Hungary–Budapest zoo).

Oocyst Structure. The following description is taken from
Pellérdy (1965). Kotlan and Pospesch (1933) gave little description except to say that the oocysts resembled those of *I. rivolta.*
Oocysts smooth, colorless, 26–34 x 18–27 μm, with thin wall,
without micropyle, residuum, or polar granule. Sporocysts ellipsoidal, 14–16 x 12 μm, without Stieda body, with residuum which
disappears in a few days. Sporozoites elongate, tapering at one
end, with a clear globule at the broad end.

Sporulation. 2–3 days (Pellérdy, 1965).

Prepatent Period. 7–8 days (Pellérdy, 1965).

Cross-Transmission Studies. Kotlan and Pospesch (1933) were
unable to infect the cat with this species.

Remarks. Pospesch was Pellérdy's name before he changed it.

Host Genus *Mephitis*

Eimeria mephitidis Andrews, 1928

(Plate 7, Fig. 39)

Type Host. Striped skunk *Mephitis mephitis.*

Location. Unknown; oocysts found in feces.
Geographic Distribution. North America (Ohio—animal supply house).
Oocyst Structure. Oocysts broadly ovoid to spherical, 17–25 x 16–22 (mean, 21 x 19) μm, with smooth, colorless wall nearly 1 μm thick, apparently composed of 2 layers, with micropyle, without residuum or polar granule. Sporocysts ovoid, 10–12 x 7–9 μm, with Stieda body and residuum. Sporozoites falcate, with one end smaller than the other, 10–14 x 4–5 μm, lying lengthwise head to tail in sporocysts, apparently without clear globule.

Eimeria voronezhensis n. sp.

(Plate 16, Fig. 86)

Synonym. Eimeria mephitidis Andrews, 1928 of Yakimoff and Matikaschwili (1932).
Type Host. Striped skunk *Mephitis mephitis.*
Location. Unknown; oocysts found in feces.
Geographic Distribution. USSR (fur farm at Voronezh).
Prevalence. Yakimoff and Matikaschwili (1932) found this species in 77% of 13 skunks on a raccoon farm in Voronezh, USSR.
Oocyst Structure. Oocysts ovoid, ellipsoidal, subspherical, or spherical, the first 3 types being 17–27 x 15–23 (mean, 23 x 20.5) μm, and the spherical ones 18–25 (mean, 22) μm in diameter, with wall said to be double-contoured or sometimes with 3 contours (drawing shows 2 layers), without micropyle, residuum, or polar granule. Sporocysts ellipsoidal, 9–12 x 7–8 μm, with a 2-layered wall, without Stieda body, with residuum. Sporozoites piriform and bent, 6–7 x 2–3 μm, illustrated without clear globule.
Cross-Transmission Studies. Yakimoff and Matikaschwili (1932) fed oocysts to a "whelp" (presumably of a dog, but possibly of a skunk or raccoon). Oocysts appeared in its feces on day 9 and persisted until day 13. These oocysts sporulated in 1 day at room temperature in 2.5% potassium bichromate solution.
Remarks. Although Yakimoff and Matikaschwili (1932) said that this species was *E. mephitidis,* it differed from that species in the absence of a micropyle and Stieda body and in sporozoite size. We are therefore giving it a new name.

Host Genus *Spilogale*

Isospora spilogales Levine and Ivens, 1964

(Plate 14, Fig. 74)

Type Host. Spotted skunk *Spilogale putorius ambarvalis.*
Location. Unknown; oocysts found in feces.
Geographic Distribution. North America (Florida).
Oocyst Structure. Oocysts somewhat ovoid, ellipsoidal, or sometimes a little asymmetrical, 29–38 x 22–28 (mean, 34 x 26) μm,
with smooth, colorless or pale grayish yellow, 1-layered wall 1–1.2
μm thick, with a tiny bleb of material adhering to the inside of the
broad end in both sporulated and unsporulated oocysts, without
micropyle, residuum, or polar granule. Sporocysts ellipsoidal, 17–
22 x 13–16 (mean, 19 x 14) μm with wall about 0.5 μm thick, without Stieda body, with residuum. Sporozoites fat sausage-shaped,
with a clear subcentral nucleus, lying at random in sporocysts.
Levine and Ivens (1964) also saw a few *Caryospora*-like oocysts
with 8 sporozoites in a single sporocyst; they were the same size as
those of *I. spilogales* and were considered to be simply abnormal
oocysts of *I. spilogales.*

Isospora sengeri Levine and Ivens, 1964

(Plate 8, Fig. 47)

Type Host. Spotted skunk *Spilogale putorius ambarvalis.*
Location. Unknown; oocysts found in feces.
Geographic Distribution. North America (Florida).
Oocyst Structure. Ellipsoidal, sometimes slightly squared at
one end, 16–23 x 12–18 (mean, 20 x 15) μm, with smooth, colorless to pale yellowish, 1-layered wall about 0.6–0.8 μm thick, without micropyle, residuum, or polar granule. In some oocysts, oocyst
wall stretched tightly around sporocysts, in a few, depressed between sporocysts. Sporocysts ellipsoidal, often lying against each
other with one side somewhat flattened, 10–14 x 8–12 (mean, 12
x 10) μm, with wall about 0.4–0.6 μm thick, without Stieda body,
with residuum. Sporozoites elongate sausage-shaped, with one end

somewhat more pointed than the other, usually lying lengthwise in sporocysts, with a clear subcentral nucleus often visible.

Host Family VIVERRIDAE

Host Genus *Genetta*

Eimeria genettae Agostinucci and Bronzini, 1955

Type Host. Genet *Genetta dongolana.*
Location. Unknown; oocysts presumably found in feces.
Geographic Distribution. Africa (Somalia—Rome [Italy] zoo).
Oocyst Structure. Slightly ellipsoidal, 20–30 x 12–25 (mean, 25 x 20) μm, with smooth, colorless, apparently double-contoured wall, with micropyle, without residuum and polar granule. Sporocysts irregularly piriform, 6–13 x 5–8 (mean, 8 x 6) μm, apparently without residuum.
Sporulation. 7–8 days at room temperature (about 18 C) in 3% potassium bichromate solution.

Host Genus *Civettictis*

Isospora viverrae Adler, 1924

(Plate 14, Fig. 75)

Type Host. African civet cat *Civettictis* (synonym, *Viverra*) *civetta.*
Location. Epithelial cells of small intestine.
Geographic Distribution. Africa (Sierra Leone).
Oocyst Structure. Ellipsoidal, 19–28 x 15–25 μm, most commonly 23 x 19 μm, with 1-layered wall, without micropyle, with residuum (or polar granule[s]) during the first division, which disappears before sporulation is complete. Sporocysts ellipsoidal, 12–15 x 8–11 μm, without Stieda body, with residuum. Sporozoites sickle-shaped, 9–11 x 3 μm, with clear globule at large end.
Sporulation. Oocysts passed in the feces contain a single sporont or 2 sporoblasts. Sporulation time thereafter 3 days.

Merogony. There may be as many as 3 generations of meronts. They reach a diameter of about 10–18 μm and contain 4–13 merozoites of variable size; they may or may not have a residuum (Adler, 1924).

Gametogony. There are about 30 times as many macrogametes as microgamonts. Both are in the small intestine epithelial cells like the meronts. The mature microgamonts are 16–25 μm in diameter and contain about 200 biflagellate microgametes about 10 μm long. Adler (1924) said that fertilization takes place inside the host cell and that he saw as many as 9 or 10 microgametes inside a single macrogamete.

Pathogenicity. According to Adler (1924) infection with this species killed 2 young civets. They passed blood and mucus in their feces. The subepithelial tissue of the infected areas was markedly hyperemic. Cells in the distal parts of the villi were heavily infected, while few cells in the basal parts of the villi were infected; in some places the distal villar epithelium had been destroyed. There was no ulceration.

Cross-Transmission Studies. Adler (1924) was unable to infect 2 cats, 3 kittens, and 2 young dogs with this species.

Remarks. Adler (1924) and others called the host *Viverra civetta,* but species of the genus *Viverra* occur only in eastern Asia. Walker et al. (1975) gave the host genus as *Civettictis* Pocock, 1915.

Host Genus *Herpestes*

Eimeria newalai Dubey and Pande, 1963

(Plate 13, Figs. 70, 71)

Type Host. Indian gray mongoose *Herpestes edwardsi* (synonym, *H. mungo*).
Location. Unknown.
Geographic Distribution. Asia (India).
Oocyst Structure. Ovoid to ellipsoidal, 15–19 x 11–17 (mean, 19 *[sic]* x 15) μm, with 2-layered wall, outer layer yellowish to orange, relatively thick, inner layer darker and thinner, without micropyle, said to be without residuum or polar granule but a "small unorganized mass was constantly seen in all the oocysts

examined." Sporocysts almost ovoid, 8–10 x 6–8 (mean, 10 x 8 [sic] μm, with Stieda body and residuum. Sporozoites comma-shaped, 8–9 x 2–3 μm, with clear globule at large end.
Sporulation. 1 day.

Eimeria pandei Patnaik and Ray, 1965 emend.
Patnaik and Ray, 1966

Synonym. Eimeria pandeii Patnaik and Ray, 1965.
Type Host. Indian gray mongoose *Herpestes edwardsi.*
Location. Unknown.
Geographic Distribution. Asia (India).
Oocyst Structure. Ovoid, 23–24 x 17–19 (mean, 23 x 19) μm, with brownish yellow wall averaging 2 μm thick but thinning at small end, without micropyle, residuum, or polar granule. Sporocysts ellipsoidal or piriform, 11 x 6 μm with residuum. Sporozoites elongate, 9.5 μm long.
Sporulation. 2 days.

Isospora dasguptai Levine, Ivens, and Healy, 1975

(Plate 11, Fig. 63)

Synonyms. Isospora rivolta (Grassi, 1879) of Knowles and Das Gupta (1931); *I. garnhami* (small form) Bray, 1954 of Dubey and Pande (1963).
Type Host. Small Indian mongoose *Herpestes auropunctatus.*
Other Host. Indian gray mongoose *H. edwardsi* (synonym, *H. mungo*).
Location. Unknown.
Geographic Distribution. Asia (India).
Oocyst Structure. Ellipsoidal or ovoid, 21 x 17 μm, without micropyle, residuum, or polar granule. Sporocysts apparently ellipsoidal. Sporozoites elongate, with one end pointed, apparently with clear globule at broad end. The above description was given by Levine, Ivens, and Healy (1975) on the basis of the text and drawing by Knowles and Das Gupta (1931), who reported this species from *H. auropunctatus.* According to Dubey and Pande

(1963), who saw it in *H. edwardsi*, the oocysts are spherical to ellipsoidal, 19–22 x 16 x 19 μm, with a 2-layered wall, without a micropyle, residuum, or polar granule, and the sporocysts are 14–17 x 8–11 μm, without a Stieda body, with a residuum.
Sporulation. 1 day (Dubey and Pande, 1963).

Isospora herpestei Levine, Ivens, and Healy, 1975

(Plate 8, Fig. 46)

Type Host. Small Indian mongoose *Herpestes auropunctatus.*
Location. Unknown; oocysts found in feces.
Geographic Distribution. North America (imported into Georgia from India).
Oocyst Structure. Subspherical to ellipsoidal, 18–24 x 13–18 (mean, 20 x 15) μm, with thin, smooth, colorless, delicate, easily collapsed wall composed of a single layer less than 0.4 μm thick, without micropyle, residuum, or polar granule. Sporocysts ellipsoidal, 11–15 x 8–11 (mean, 13 x 9) μm, with slightly yellowish wall about 0.5 μm thick, without Stieda body, with residuum. Sporozoites sausage-shaped, with both ends of equal diameter, lying at random in sporocysts, without clear globule.
Pathogenicity. Levine, Ivens, and Healy (1975) thought that this species caused diarrhea.

Isospora mungoi Levine, Ivens, and Healy, 1975

(Plate 11, Fig. 60)

Synonym. Large *Isospora garnhami* Bray, 1954 of Dubey and Pande (1963).
Type Host. Indian gray mongoose *Herpestes edwardsi* (synonym, *H. mungo*).
Location. Unknown; oocysts found in feces.
Geographic Distribution. Asia (India).
Oocyst Structure. Ellipsoidal, with rounded ends, 27–34 x 23–27 (mean, 34 x 27) *[sic]*, μm, with 2-layered wall 1–2 μm thick, outer layer the thicker, light yellowish green to pale yellow, inner

layer darker, without micropyle, residuum, or polar granule. Sporocysts ellipsoidal, 19–21 x 12–24 (mean, 21 x 14) [sic] μm, without
Stieda body, with residuum. Sporozoites sausage-shaped, with one
end narrower than the other, with clear globule at broad end. The
above description is from Dubey and Pande (1963).

Sporulation. 1 day.

Isospora pellerdyi Dubey and Pande, 1964

(Plate 13, Fig. 73)

Synonym. Isospora knowlesi Dubey and Pande, 1963.
Type Host. Indian gray mongoose *Herpestes edwardsi* (synonym,
H. mungo).
Location. Unknown; oocysts found in feces.
Geographic Distribution. Asia (India).
Oocyst Structure. Ovoid, 27–30 x 20–25 (mean, 28 x 23) μm,
with 2-layered wall, outer layer the thicker, yellowish to orange,
inner layer darker, without micropyle, polar granule, or residuum.
Sporocysts ellipsoidal, 17–19 x 12–24 (mean, 15 x 11) [sic] μm,
without Stieda body, with residuum. Sporozoites elongate, with
one end pointed, 13–15 x 2–3 μm.
Sporulation. 1–2 days.

Isospora ichneumonis Levine, Ivens, and Healy, 1975

Synonym. Isospora rivolta (Grassi, 1879) of Balozet (1933).
Type Host. Egyptian mongoose *Herpestes ichneumon.*
Location. Unknown; oocysts found in feces.
Geographic Distribution. Africa (Tunisia).
Oocyst Structure. Subspherical, 19–26 x 16–20 (mean, 22 x 19)
μm, with double-contoured, colorless, transparent, easily deformed
wall, with poorly visible micropyle. Sporocysts rounded, slightly
ovoid, with a wall thinner than that of the oocyst, with residuum.
Sporozoites banana-shaped, parallel. The above description is from
Balozet (1933).
Sporulation. 4 days at 20 C on filter paper impregnated with
potassium bichromate solution (Balozet, 1933).

Cross-Transmission Studies. Balozet (1933) claimed to have infected a puppy with this species, but it is impossible to determine whether he did or whether the oocysts passed by the puppy were a dog species.

Host Genus *Helogale*

Isospora garnhami Bray, 1954

(Plate 11, Fig. 62)

Type Host. Dwarf mongoose *Helogale undulata rufula.*
Other Host. Cusimanse (mongoose) *Crossarchus obscurus.*
Location. Ileum and colon.
Geographic Distribution. Africa (Kenya—*H. undulata;* Liberia— *C. obscurus*).
Oocyst Structure. The oocysts in *H. undulata* were described by Bray (1954) as ellipsoidal, 26–32 x 22–28 (mean, 29 x 24.5) μm, with a 2-layered wall, outer layer thin and fragile, inner layer tough and elastic, without micropyle, residuum, or polar granule. Sporocysts spherical to ellipsoidal, 12–15 x 11–12 (mean, 14 x 11.5 μm, without Stieda body, with residuum. Sporozoites sausage-shaped, slightly tapered at one end, 14 x 2 μm, with clear globule. The oocysts in *C. obscurus* were described by Bray (1954) as 22–30 x 18–22 (mean, 27 x 20) μm, with a 2-layered wall. He said that these oocysts were indistinguishable in general structure from those of *I. garnhami* except for one feature: when the oocysts were released into the lumen of the intestine, their outer wall was ellipsoidal and their inner wall flattened at both ends, but by the time they were voided both walls were ellipsoidal.
Sporulation. The oocysts from *H. undulata* sporulated in 3 days at room temperature (22 C) in 0.25% chromic acid or 4% sodium bichromate solution.
Merogony. Bray (1954) found only 3 meronts in *H. undulata.* They were in a single patch of tissue from the ileum and contained 12–24 merozoites.
Gametogony. Bray (1954) found all stages of gametogony in the lower small intestine and large intestine of *H. undulata.* They were in the villar epithelial cells above the host cell nucleus. The

macrogametes bulged into the subepithelial tissue when mature, and averaged 24 x 22 μm in fixed tissue. The microgamonts averaged 23 x 22 μm in fixed tissue and contained several hundred microgametes.

Cross-Transmission Studies. Bray (1954) was unable to infect a ferret *Mustela putorius furo* or a kitten with this species from *H. undulata.*

Isospora hoarei Bray, 1954

(Plate 8, Fig. 48)

Type Host. Dwarf mongoose *Helogale undulata rufula.*
Location. Epithelial cells of duodenum and jejunum.
Geographic Distribution. Africa (Kenya).
Oocyst Structure. Ellipsoidal, 16–19 x 13–17 (mean, 17 x 15) μm, with 2-layered wall, outer layer colorless, inner layer dark, more fragile than *I. garnhami* wall, without micropyle, residuum, or polar granule. Sporocysts almost spherical, 9–10 x 8–9 (mean, 9 x 9) μm, without Stieda body, with large residuum. Sporozoites sausage-shaped, slightly tapering at one end, 7 x 2 μm, apparently without clear globule.
Sporulation. 4 days at room temperature (22 C) in 0.25% chromic acid or 4% sodium bichromate solution.
Gametogony. Bray (1954) found no meronts, but did find mature macrogametes 16 x 14 μm and mature microgamonts 17 x 16 μm (both fixed tissue). He counted 117 microgametes in 1 microgamont.
Cross-Transmission Studies. Bray (1954) was unable to infect a ferret *Mustela putorius furo* or a kitten with this species from *H. undulata.*

Sarcocystis sp. Viljoen, 1921

Definitive Host. Unknown.
Intermediate Host. Dwarf mongoose *Helogale parvula.*
Location. Striated muscles.

Geographic Distribution. Africa (South Africa).
Merogony. Only sarcocysts known.
Remarks. Viljoen (1921) report not seen; the above is from Neitz (1965).

Host Genus *Mungos*

Sarcocystis sp. Viljoen, 1921

Definitive Host. Unknown.
Intermediate Host. Striped mongoose *Mungos mungo.*
Location. Striated muscles.
Geographic Distribution. Africa (South Africa).
Merogony. Only sarcocysts known.
Remarks. Viljoen (1921) report not seen; the above is from Neitz (1965).

Host Genus *Crossarchus*

See *Isospora garnhami* under host genus *Helogale* above.

Host Family HYAENIDAE

Host Genus *Hyaena*

Isospora levinei Dubey, 1963

(Plate 14, Fig. 77)

Type Host. Striped hyena *Hyaena hyaena* (synonym, *H. striata*).
Location. Unknown.
Geographic Distribution. Asia (India).
Oocyst Structure. Ovoid, 23–29 x 22–26 (mean, 27 x 25) μm, with 2-layered wall 1–2 μm thick, outer layer the thicker, smooth, yellowish green, inner layer darker with shining inner contour, without micropyle, residuum, or polar granule. Sporocysts ellipsoidal, 16–18 x 11–14 (mean, 18 x 14) μm, without Stieda body,

with residuum. Sporozoites 10–14 x 3–4 μm, with one end rounded and the other pointed, apparently without clear globule.
Sporulation. 1 day.

Host Family FELIDAE

Host Genus *Lynx*

Eimeria lyncis Anpilogova and Sokov, 1973

Type Host. Central Asian lynx *Lynx lynx isabellina.*
Location. Unknown; oocysts found in feces.
Geographic Distribution. USSR (Tadzhikistan).
Oocyst Structure. Elongate ellipsoidal, 30–38 x 19–27 (mean, 33 x 23) μm, with rough, yellowish, 2-layered wall, with micropyle in inner layer. Sporocysts ellipsoidal, with clearly visible Stieda body and residuum. No illustration given.

Eimeria tadzhikistanica Anpilogova and Sokov, 1973

Type Host. Central Asian lynx *Lynx lynx isabellina.*
Location. Unknown; oocysts found in feces.
Geographic Distribution. USSR (Tadzhikistan).
Oocyst Structure. Ovoid, colorless, smooth, 24–32 x 19–27 (mean, 31 x 23.5) μm, with 2-layered wall, without micropyle or residuum. Sporocysts ellipsoidal, with residuum. No illustration given.

Eimeria sp. Anpilogova and Sokov, 1973

Type Host. Central Asian lynx *Lynx lynx isabellina.*
Location. Unknown; single oocyst found in feces.
Geographic Distribution. USSR (Tadzhikistan).
Oocyst Structure. Elongate ovoid, truncate at micropylar end, 40.5 x 27 μm. Sporocysts 13.5 x 11 μm. No illustration given.

Isospora lyncis n. sp.

Synonym. Isospora felis (Wasielewski, 1904) Wenyon, 1923 from lynx of Triffitt (1927).

Type Host. Lynx *Lynx* sp.

Location. Unknown; oocysts found in rectal contents.

Geographic Distribution. Europe (London zoo).

Oocyst Structure. The following description was given by Triffitt (1927). Oocysts oval, 40–47 x 28–37 μm, with 3-layered wall, the middle one slightly yellowish and 0.75 μm thick, with micropyle 4–5 μm in diameter. Sporocysts ovoid with double-contoured wall, 20–33 x 14–18 μm, with residuum. Sporozoites elongate, about 15 μm long, with large greenish clear globule at broad end and sometimes similar but smaller one at narrow end.

Sporulation. 7–9 days at room temperature.

Remarks. Triffitt (1927) said that she found oocysts of this species in the rectal contents of a lynx that died in the Zoological Gardens of London. However, she did not say whether it was the European lynx *Lynx lynx* or the Canada lynx *L. canadensis.* Further, although she said that its oocysts were identical with those of *I. felis,* they had a distinct micropyle that *I. felis* lacks and were therefore clearly different. We are therefore giving this form a new name.

Host Genus *Felis*

Eimeria cati Yakimoff, 1933

(Plate 14, Fig. 76; Plate 15, Fig. 78)

Type Host. Domestic cat *Felis catus.*

Other Hosts. European wild cat *F. silvestris;* Indian jungle cat *F. chaus.*

Location. Unknown; oocysts found in feces.

Geographic Distribution. USSR (Leningrad, Azerbaidzhan, Transcaucasia), Asia (Iraq, India).

Prevalence. Mirza (1970) reported this species from 3% of 30 domestic cats in Iraq.

Oocyst Structure. The oocysts described by Yakimoff (1933) from the domestic cat in Leningrad and Azerbaidzhan were ellipsoidal, ovoid, or spherical, 18-23 x 14-20 (mean, 21 x 17) μm (spherical oocysts, 16-22 μm in diameter with a mean of 18 μm), without a micropyle or residuum, with a polar granule, with ovoid sporocysts 11 x 6 μm, presumably without a Stieda body, with a residuum. The oocysts found by Gousseff (1933) in a wild cat in Transcaucasia were spherical, ovoid, or ellipsoidal, 14-27 x 11-22 μm, most frequently 18 x 16 μm (spherical oocysts, 12-16 μm, most frequently 16 μm, in diameter), colorless, apparently with a 2-layered wall, the outer layer being darker than the inner, without a micropyle or residuum, with a polar granule, with sporocysts 5-8 x 4-5 μm, most frequently 7 x 4 μm, with a residuum but apparently without a Stieda body. The oocysts found by Dubey and Pande (1963a) in the Indian jungle cat in India were subspherical to spherical, 19-25 x 19-22 (mean, 21 x 19) μm, with a 2-layered wall 1-2 μm thick, outer layer thicker and grayish, inner layer darker, with a shining inner contour, without micropyle or residuum, with polar granule, with ellipsoidal sporocysts 10-11 x 7 μm, without a Stieda body, with a residuum, with elongate sporozoites apparently with a clear globule.

Sporulation. According to Dubey and Pande (1963a), the sporulation time of the *F. chaus* oocysts was 8 days.

Remarks. This may or may not be an authentic parasite of cats. Perhaps it is a pseudoparasite.

Eimeria felina Nieschulz, 1924

(Plate 11, Figs. 61, 64)

Type Host. Domestic cat *Felis catus.*

Other Hosts. Indian jungle cat *F. chaus;* lion *Leo* (synonym, *Panthera*) *leo.*

Location. Unknown; oocysts found in feces.

Geographic Distribution. Europe (Belgium, Netherlands), USSR (Leningrad zoo), Asia (India).

Prevalence. Fameree and Cotteleer (1976) reported this species from 5 of 252 cats in Belgium.

Oocyst Structure. The oocysts described from the domestic cat by Nieschulz (1924) in the Netherlands were long, ellipsoidal to cylindrical, 21-26 x 13-17 (mean, 24 x 14.5) μm, very bright, almost colorless, without a discernible micropyle but with one end somewhat flattened and with a somewhat thinner wall than the rest, with residuum, apparently without polar granule, with ovoid sporocysts apparently with Stieda body, with residuum, with elongate comma-shaped sporozoites lying lengthwise head to tail in sporocysts and containing a large clear globule at the broad end. The oocysts found by Rastegaieff (1929, 1930) in a lion in the Leningrad zoo were ovoid and 21 x 16 μm without a visible micropyle. The oocysts seen by Dubey and Pande (1963) in an Indian jungle cat in India were ovoid ellipsoidal, 15-19 x 11-17 (mean, 19 x 15) μm, with a 2-layered wall, outer layer thicker and yellowish to orange, inner layer darker with a shining inner contour, without micropyle, residuum, or polar granule but "an unorganized mass was constantly seen," with nearly ovoid sporocysts 8-10 x 6-8 (mean, 10 x 8) μm, with a prominent Stieda body and a residuum, with large, comma-shaped sporozoites 8-9 x 2-3 μm, lying lengthwise head to tail in the sporocysts, the presence of a clear globule not determinable.

Sporulation. The oocysts found by Dubey and Pande (1963) in *F. chaus* sporulated in 1 day.

Remarks. This may or may not be an authentic parasite of cats. Perhaps it is a pseudoparasite.

Eimeria chaus Ryšavý, 1954

Type Host. Jungle cat *Felis chaus.*
Location. Unknown; oocysts found in feces.
Geographic Distribution. Europe (Prague zoo).
Oocyst Structure. Oocysts broad oval, 18-24 x 14-22 μm, with smooth, 1-layered wall 1.8 μm thick, tapering at one end, without micropyle. Sporocysts ellipsoidal, 9-11 x 4-6 μm, with a residuum.

Remarks. This may or may not be a genuine parasite of the jungle cat. Perhaps it is a pseudoparasite.

Eimeria hammondi Dubey and Pande, 1963

(Plate 9, Fig. 52)

Type Host. Indian jungle cat *Felis chaus.*
Location. Unknown; oocysts found in feces.
Geographic Distribution. Asia (India).
Oocyst Structure. Ellipsoidal, 24–29 x 19–22 μm, with a 2-layered wall 1–2 μm thick, outer layer the thicker and grayish, inner layer darker, with a shining inner contour, without micropyle, residuum, or polar granule. Sporocysts broadly ovoid, 11–14 x 8–10 (mean, 12 x 8) μm, with very small Stieda body, with residuum. Sporozoites comma-shaped, 8–10 x 2–3 μm, with central nucleus and clear globule at broad end.
Sporulation. 6 days.
Remarks. This may or may not be a genuine parasite of the jungle cat. Perhaps it is a pseudoparasite.

Eimeria mathurai Dubey and Pande, 1963

(Plate 12, Figs. 66, 67)

Type Host. Indian jungle cat *Felis chaus.*
Location. Unknown; oocysts found in feces.
Geographic Distribution. Asia (India).
Oocyst Structure. Ellipsoidal or broadly spindle-shaped, 20–28 x 16–20 (mean, 22 x 19) μm, with 2-layered wall, outer layer thicker, smooth, light yellowish green to pale yellow, inner layer darker, with shining inner contour, without micropyle or residuum, with polar granule. Sporocysts broadly ovoid, 11–13 x 7–9 (mean, 11 x 8) μm, with prominent Stieda body, with residuum. Sporozoites comma-shaped, 8–10 x 2–3 μm, lying lengthwise head to tail in sporocysts, apparently with or without a clear globule at the broad end.
Sporulation. 6 days.
Remarks. This may or may not be a genuine parasite of the jungle cat.

Sarcocystis sp. (Dubey and Pande, 1963) nov. comb.

Synonym. Cryptosporodial [sic] coccidial bodies of Dubey and Pande, 1963.
Type Definitive Host. Jungle cat *Felis chaus.*
Intermediate Host. Unknown.
Location. Feces.
Geographic Distribution. Asia (India).
Oocyst Structure. Oocysts unknown. Mature sporocysts found in feces. They were ellipsoidal, 10–12 x 7–8 (mean, 11 x 7) μm, with a 2-layered wall, outer layer thick, pale yellow, inner layer green, with a residuum. They contained 4 sporozoites, 7–9 x 1–2 μm, with one end rounded and the other pointed, with a clear globule at the broad end.

Isospora felis Wenyon, 1923

(Plate 12, Fig. 69)

Synonyms. Diplospora bigemina of Wasielewski (1904) in part; *Isospora bigemina* of Swellengrebel (1914); *I. rivolta* Dobell and O'Connor, 1921; *I. cati* Marotel, 1921; *Lucetina felis* (Wenyon, 1923) Henry and Leblois, 1926; *I. felis* var. *servalis* Mackinnon and Dibb, 1938; *Cystoisospora felis* (Wenyon, 1923) Frenkel, 1977; *Levinea felis* (Wenyon, 1923) Dubey, 1977.
Type Definitive Host. Domestic cat *Felis catus.*
Other Definitive Hosts. European wild cat *F. silvestris* (Prasad, 1961); ocelot *F. pardalis* (Lainson, 1968); serval *F. serval* (Mackinnon and Dibb, 1938) tiger *Leo tigris* (Yakimoff et al., 1933; Prasad, 1961); lion *L. leo* (Yakimoff et al., 1933); jaguar *L. onca* (Barreto and de Almeida, 1937); lynx *Lynx lynx* (Yakimoff et al., 1933).
Transport Hosts. All the following are experimental: House mouse *Mus musculus* (Frenkel and Dubey, 1972; Christie, Dubey, and Pappas, 1976; Dubey and Streitel, 1976), Norway rat *Rattus norvegicus* (Frenkel and Dubey, 1972), golden hamster *Mesocricetus auratus* (Frenkel and Dubey, 1972), cat *F. catus* (Dubey and Frenkel, 1972), dog *Canis familiaris* (Dubey, 1975), and ox *Bos taurus* (Fayer and Frenkel, 1979).

Location. Small intestine, sometimes cecum, occasionally colon.
Geographic Distribution. Worldwide.
Prevalence. This species is probably the commonest coccidium
of cats. It was found in 13% of 130 cats in Illinois by Shah (1970),
in 23% of 217 cats in Illinois by Guterbock and Levine (1977), in
25% of 237 cats in New Jersey by Burrows (1968), in 27% of 757
stray cats in New Jersey by Burrows and Hunt (1970), in 17% of
510 domiciled cats in Kansas City by Dubey (1973) (20% of 303
kittens 4.5–10 weeks old, 22% of 87 kittens 11–26 weeks old, and
6% of 126 cats 6 months or more old); in 6% of 1,000 cats in Ohio
by Christie, Dubey, and Pappas (1976), in 27% of 11 cats in Mexico
City by Davalos and Briseño (1960), in 39% of 125 cats in Brazil
by Nery-Guimaraes and Lage (1973), in 13% of 100 cats in Valdivia,
Chile, by Torres, Hott, and Boehmwald (1972), in 14% of 50 cats
in the Netherlands by Nieschulz (1925), in 5% of 252 cats in Belgium by Fameree and Cotteleer (1976), in 10% of 40 cats in Portugal
by da Cruz, de Sousa, and Cabral (1952), in 7% of 106 stray cats
in Leningrad by Machul'skii and Timofeev (1940), in 14% of 50
cats in Madras, India, by Alwar and Lalitha (1958), in 12% of 200
cats in Japan by Tomimura (1957), in 10% of 100 cats in Japan by
Iseki et al. (1974), in 9% of 446 cats in Japan by Ito et al. (1974),
in 17% of 30 cats in Iraq by Mirza (1970), and in 18% of 50 stray
cats in Sydney, Australia, by Bearup (1960).

Oocyst Structure. The following description, based primarily
on Wenyon (1923) and Shah (1969, 1970), is taken from Levine
(1973). Oocysts ovoid, 32–53 x 26–43 (mean, 42–43 x 31–33) μm,
with a smooth, very pale yellow to pale brown, 1-layered wall
about 1.3 μm thick apparently lined by a thin membrane, without
micropyle or residuum, generally without polar granule. Sporocysts
ellipsoidal, 20–27 x 17–22 (mean, 23 x 18) μm, with a smooth,
colorless wall 0.4 μm thick, without a Stieda body, with a residuum.
Sporozoites sausage-shaped with one end slightly narrowed, 10–15
μm long, generally lying lengthwise in sporocysts, with a subcentral
clear globule. Shah (1969, 1970) found a very few *Caryospora*-like
oocysts with a single sporocyst containing 8 sporozoites in 1 cat
infected with *I. felis.*

Sporulation. 2 days or less. Shah (1969, 1970a) found that it
was 40 hours at 20 C in 2.5% potassium bichromate solution, 24

hours at 25 C, 12 hours at 30 C, and 8 hours at 38 C; no sporulation occurred at 45 and 50 C, and the oocysts died.

Merogony. Hitchcock (1955) and Lickfeld (1959) described the life cycle of *I. felis,* but their descriptions differed in many respects. Shah (1969, 1971) studied the life cycle carefully, using a strain of *I. felis* derived from a single oocyst; he largely confirmed Lickfeld's account. The endogenous stages occur in the epithelial cells of the distal parts of the villi of the ileum and occasionally duodenum and jejunum. All stages lie above the host cell nucleus. There are 3 asexual generations. The first-generation meronts are 11-30 x 10-23 μm when mature and contain 16-17 banana-shaped merozoites, 11-15 x 3-5 μm. They become mature in 96 or sometimes 120 hours. The first-generation merozoites enter new host cells, round up, and form second-generation meronts. These form within themselves 2-10 spindle-shaped bodies (second-generation merozoites) resembling first-generation merozoites. They are uninucleate 120 hours after inoculation, but by 144 hours they are larger, multinucleate, and some of them have lost their elongate shape and become ovoid. These are third-generation meronts. They are 12-16 x 4-5 (mean, 14 x 5) μm. Each forms up to 6 banana-shaped merozoites 6-8 x 1-2 (mean, 7 x 2) μm. These third-generation merozoites contain a central nucleus with a prominent nucleolus; Lickfeld called them micromerozoites because they are considerably smaller than the first-generation merozoites. They are formed within the "cyst" of the second-generation meront; they are not all formed at the same time, so that "cysts" may be found containing both fully formed third-generation merozoites and second-generation merozoites. Mature "cysts" containing only third-generation merozoites are found 6-9 days after inoculation and are most abundant 7 days after inoculation. They contain 36-70 or more merozoites, which are not arranged in any particular fashion. Thus, the third-generation meronts and merozoites develop within the same host cell and parasitophorous vacuole as do the second-generation meronts and merozoites.

Hypnozoites in the lymph nodes of transport hosts have a crystalloid body like that of *I. canis* (Roberts, Mahrt, and Hammond, 1972). Mehlhorn and Markus (1976) described their fine structure in the mesenteric lymph nodes of the mouse 25 days after inoculation. The parasites occur singly in parasitophorous vacuoles. They resemble sporozoites, but have a large crystalloid body posterior to the nucleus. There is no "cyst" wall.

Gametogony. The third-generation merozoites break out of their host cell and enter other epithelial cells. They round up to form gamonts, which first appear 6 days after inoculation and reach their maximum number at 8-9 days. Mature microgamonts are 24-72 x 18-32 µm and contain a central residuum and a large number of microgametes. The latter are 5-7 x 0.8 µm and have 2 posteriorly directed flagella. The macrogametes grow without further dividing. They are 16-22 x 8-13 µm when mature and have a large central nucleus and nucleolus. After fertilization by a microgamete, they turn into oocysts, which are passed in the feces.

According to Walton (1959), the haploid number of chromosomes in *I. felis* is 2. Wenyon (1923) said that there were about 8-10 chromosomes. Lickfeld (1959) described a cryptomitotic type of schizogony, but distinguished no chromosomes.

Prepatent Period. 7 days (Tomimura, 1957; Chessum, 1972; Dubey, 1976); 7-8 days (Lickfeld, 1959; Shah, 1971); 7-11 days (Dubey and Frenkel, 1972a); 8-10 days (Frenkel and Dubey, 1972); 7-10 days (mean, 8.5 days) (Dubey and Streitel, 1976).

Patent Period. 16-20 days (Tomimura, 1957); 16 days (Chessum, 1972); 10-11 days with peak oocyst production on day 6 (Shah, 1971).

Pathogenicity. This species is slightly, if at all, pathogenic under normal circumstances. None of 18 kittens 4-9 weeks old fed 100,000 sporulated oocysts by Hitchcock (1955) showed signs of disease. Shah (1969) saw no clinical signs of illness in kittens 1.5-2 months old fed as many as 150,000 oocysts. However, Tomimura (1957) observed a watery mucous diarrhea, anorexia, slight rise in temperature, failure to digest food, depression, weakness, anemia, emaciation, and loss of weight in artificially infected cats. The disease was usually self-limiting, but heavy infections were fatal. The duration of signs and the severity of infection varied with the number of oocysts fed, but Tomimura (1957) did not tell how many he gave or how many cats (or kittens?) were used. Andrews (1926) observed enteritis, emaciation, weakness, depression, dysentery, and even death in kittens experimentally infected with *I. felis.* Hitchcock (1955) thought that these signs and deaths in the kittens might well have been due to feline distemper. Dubey and Streitel (1976) did not consider *I. felis* to be a serious feline pathogen. The gross pathologic lesions reported have been hemorrhagic enteritis, frequently with ulceration, thickened mucosa, and epithelial desquamation. Tomimura (1957) said that the erythrocyte numbers decreased and the leukocyte numbers increased beginning about 3 days after inoculation. The lymphocytes and eosinophils increased, and a few plasma cells appeared in the peripheral blood.

Immunity. Cats that have recovered from *I. felis* infections are resistant to reinfection.

Chessum (1972) found that no cross-immunity existed between *I. felis, I. rivolta,* and *Toxoplasma gondii.*

Cross-Transmission Studies. Patent infections do not occur in dogs fed *I. felis* oocysts from cats (Neméseri, 1959; 1960; Shah, 1969, 1970; Dubey, Miller, and Frenkel, 1970; da Rocha and Lopes, 1971; Nery-Guimaraes and Lage, 1973; Dubey, 1975; Guterbock and Levine, 1977) or in dogs fed occultly infected mice (Guterbock and Levine, 1977). Kotlan and Pospesch (1933) were unable to produce a patent infection in a young badger *Meles meles.* Nukerbaeva and Svanbaev (1974) were unable to produce patent infections in *Vulpes vulpes, Alopex lagopus,* or the mink with oocysts from the cat. However, as stated above, occult infections of various parenteral organs whose presence is revealed by the feeding of tissues to cats occur in the cat, dog, house mouse, Norway rat, and golden hamster.

Dubey and Frenkel (1972) fed kittens sporocysts of *I. felis* and then killed them at intervals and fed suspensions of their extraintestinal tissues to indicator kittens less than a day old. If the indicator kittens produced oocysts with a regular prepatent period, they concluded that coccidian stages were present in the extraintestinal tissues of the donor cats. They found that the liver and spleen were positive for *I. felis* for 45 days after inoculation, the mesenteric lymph nodes for 104 days, the lungs for 10 days, and the brain and muscle for 12 days. The prepatent period was 4–8 days as opposed to 7–11 days after consuming oocysts. They found distinct coccidian stages unlike those in the gut either alone or in groups of 2–15 in the lymphoreticular cells of mesenteric lymph nodes of 2 kittens infected for 2–4 days. They recommended touch smears rather than sections to find the extraintestinal stages.

Frenkel and Dubey (1972) found that *I. felis* produced nonpatent infections in mice and that oocysts appeared in newborn kittens after the mice had been fed to them. The prepatent period between the time the mice were fed to the kittens and the time oocysts appeared in the kittens' feces was 5–6 days; the time was 8–10 days in kittens fed oocysts. The coccidia were present in the lungs, livers, spleens, and mesenteric lymph nodes of the mice for at least 67 days. The merozoites in the mesenteric lymph nodes were 7–10 x 6–8 μm during multiplication. The "ensheathed" merozoites were 13–26 x 6–11 μm, and the "sheath" was 21–35 x 7–14 μm. The sporozoites that the mice ingested were 13 x 3–4 μm. Dubey and Streitel (1976), too, found that cats could be infected with *I. felis* by feeding them homogenized mice that had been fed oocysts 5 days before. The prepatent period in these cats was 5–9 (mean, 6.3) days; the time was 7–10 (mean, 8.5) days for kittens fed oocysts.

Frenkel and Dubey (1972) also found that *I. felis* produced nonpatent infections in Norway (laboratory) rats and golden hamsters. The organisms were in these animals' lungs, livers, spleens, and/or lymph nodes for at least 10 days, and newborn kittens were infected by feeding them these organs, but no oocysts were ever found in the rodents.

Dubey (1975) found that gnotobiotic 6-month-old dogs fed *I. felis* oocysts from cats did not shed oocysts during the next 15 days, but that they nevertheless had occult infections, since cats shed oocysts after they had been fed the dogs' extraintestinal organs.

Mehlhorn and Markus (1976) described the fine structure of *I. felis* in the mesenteric lymph nodes of the mouse 25 days after inoculation. The parasites occur singly in the parasitophorous vacuoles. They resemble sporozoites, but have a large crystalloid body posterior to the nucleus (which is comparable to the clear globule of sporozoites). There is no cyst wall. Mehlhorn and Markus (1976) considered these extraintestinal forms to be "waiting stages." There is no evidence whether they are simply enlarged sporozoites or are the result of multiplication.

There is no evidence that *I. felis* is transmitted transplacentally from mother cats to kittens (Dubey, 1977).

Gametony. The third-generation merozoites break out of their host cell and enter other epithelial cells. They round up to form gamonts, which first appear 6 days after inoculation and reach their maximum number at 8–9 days.

Cultivation. Fayer and Thompson (1974) cultivated *I. felis* in monolayer cultures of human embryonic intestine, esophageal epithelium, amnion, lung, and epithelioid carcinoma of the cervix (HeLa) cells as well as in cells from bovine embryonic trachea, canine kidney, and chick kidney. Pairs of daughter organisms developed by endodyogeny and possibly schizogony. The clear globule in the sporozoites disappeared and then reappeared in the daughter cells later, leading them to conclude that it had been formed *de novo.*

They started with excysted sporozoites and saw pairs of daughter organisms at 2 days and the greatest number 4–7 days after inoculation in the human cell types. They did not obtain oocysts.

Remarks. Mackinnon and Dibb (1938) found oocysts 26–33 x 22–27 μm in which sporocysts 13 μm in mean length developed in the feces of a serval *F. serval* in the London zoo. The oocysts were apparently partially sporulated when passed. They said that the "size of the oocysts agrees pretty well with that of *Isospora felis* as given by Wenyon (1926); but it must be admitted that the sporocyst is too small for that species." However, both the oocysts and sporocysts were smaller than those of *I. felis* and larger than those of *I. rivolta.* It is likely that this is a separate species, but in the absence of more information we do not feel justified in stating this positively.

Powell and McCarley (1975) found *Sarcocystis* sp. meronts resembling those of *S. muris* and visible to the naked eye in laboratory mice *Mus musculus* in Iowa. They fed the mice to cats, and the cats passed oocysts resembling those of *I. felis.* These oocysts in turn produced *Sarcocystis* infection in mice, and the new sarcocysts produced *I. felis*-like oocysts in cats. Their paper should be consulted for details. Later, Powell and Last (1977) said that the original mice had a mixed infection of *I. felis* and *S. muris.*

Isospora rivolta (Grassi, 1879) Wenyon, 1923

(Plate 12, Fig. 65)

Synonyms. Coccidium rivolta Grassi, 1879; *Diplospora bigemina*

Wasielewski, 1904 in part; *Isospora rivoltae* Dobell, 1919; *Isospora rivoltai* (Grassi, 1879) *auctores; Lucetina rivolta* (Grassi, 1879) Henry and Leblois, 1926; *Isospora novocati* Pellérdy, 1974; *Cystoisospora rivolta* (Grassi, 1879) Frenkel, 1977; *Levinea rivolta* (Grassi, 1879) Dubey, 1977.

Type Definitive Host. Domestic cat *Felis catus.*

Other Definitive Hosts. Wild cat (presumably *F. silvestris*) in Transcaucasia (Gousseff, 1933a); jungle cat *F. chaus* in Leningrad zoo (Yakimoff et al., 1933) and India (Dubey and Pande, 1963); tiger *Leo tigris* and leopard *L. pardus* in Leningrad zoo (Yakimoff et al., 1933).

Transport Hosts. All the following are experimental: house mouse *Mus musculus* (Frenkel and Dubey, 1972; Dubey and Streitel, 1976; Dubey, 1979), Norway rat *Rattus norvegicus* (Frenkel and Dubey, 1972), golden hamster *Mesocricetus auratus* (Frenkel and Dubey, 1972); domestic cat *F. catus* (Dubey and Frenkel, 1972; Dubey and Streitel, 1976; Dubey, 1979); and ox *Bos taurus* (Fayer and Frenkel, 1979). Dubey (1979) found that the hypnozoites in the mouse mesenteric lymph nodes were 6–8 x 4–6 (mean, 7 x 5) µm on day 1 and had grown to 11–17 x 6–9 (mean, 13 x 7) µm and were enclosed in a sheath by day 31; they did not multiply.

Location. Small intestine, cecum, and colon.

Geographic Distribution. Worldwide.

Prevalence. Common. According to Levine (1973) it was found in 3%–16% (mean, 12%) of cats in 4 surveys in the United States and in 0.9%–16% in 2 surveys in other countries. More specifically, it was found in 16% of 757 stray cats in New Jersey by Burrows and Hunt (1970), in 3% of 130 cats in Illinois by Shah (1970), in 24% of 217 cats in Illinois by Guterbock and Levine (1977), in 9% of 510 domiciled cats in Kansas City by Dubey (1973), in 7% of 100 cats in Valdivia, Chile, by Torres, Hott, and Boehmwald (1972), in 30% of 125 cats in Brazil by Nery-Guimaraes and Lage (1973), in 12% of 50 cats in Utrecht, Holland, by Nieschulz (1925), in 1% of 252 cats in Belgium by Fameree and Cotteleer (1976), in 4% of 106 stray cats in Leningrad, USSR, by Machul'skii and Timofeev (1940), in 4% of 50 cats in Madras, India, by Alwar and Lalitha (1958), in 4% of 200 cats in Japan by Tomimura (1957), in 0.5% of 446 cats in Japan by Ito et al. (1974), in 20% of 30 cats in Iraq by Mirza (1970), and in 16% of 50 cats in Sydney, Australia, by Bearup (1960).

Oocyst Structure. The following description is taken from Levine (1973) based on Shah (1969, 1970). The sporulated oocysts are ellipsoidal to somewhat ovoid, 21–28 x 18–23 (mean, 25 x 21) µm (23–29 x 20–26 [mean, 25 x 23] µm according to Dubey, 1979), with a smooth, colorless to very pale brown, 1-layered wall about 0.5 µm thick, apparently lined by a very thin membrane, without a micropyle, residuum, or polar granule. Sporocysts broadly ellipsoidal, occasionally flattened on one side, 14–16 x 10–13 (mean, 15 x 12) µm, with a smooth, colorless wall about 0.3 µm thick, without Stieda body, with residuum. Sporozoites banana-shaped, usually oriented more or less lengthwise in sporocysts, with a subcentral clear globule.

The wall of a few oocysts seen by Shah (1969, 1970) had collapsed around the sporocysts to form dumbbell-shaped doublets after sporulation. Upon refrigeration for several months, there was a high percentage of such doublets. They had a very thin oocyst wall, and the sporocysts were then about 17 x 10 µm, with a much diminished residuum.

Sporulation. 1–2 days at room temperature (Pellérdy, 1974); 1 day at 22–26 C (Dubey, 1979). Sporulation time of the form from *F. chaus* was 1 day (Dubey and Pande, 1963).

Merogony. Nery-Guimaraes and Lage (1973) said that the mature meronts averaged 10 x 6 µm and contained about 24 merozoites, 6–8 x 1–1.5 µm.

Dubey (1979) described merogony in the cat in detail. Asexual stages are in the epithelial cells of the villi and glands of Lieberkuhn of the small intestine, cecum, and colon. There are at least 3 structurally different meronts. The earliest meronts (type I) were seen 12–48 hours after feeding of mice containing hypnozoites. They were 6–13 x 3–6 (mean, 8.5 x 5) µm in sections and contained 2–8 merozoites; these merozoites had been produced by binary fission or endodyogeny. Type II meronts occurred 48–172 hours after inoculation. These meronts were in parasitophorous vacuoles, 9–18 x 9–13 (mean, 13 x 10) µm in sections that contained 1–5 merozoite-shaped meronts, 7–13 x 3–5 µm in sections or 10–16 x 5–6 µm in smears, containing up to 8 nuclei. Type III meronts occurred 72–

192 hours after inoculation. Most were in the ileum. These meronts were 7–24 x 4–21 (mean, 14 x 11) μm in sections and contained 2–30 slender merozoites, 5.5 x 1 μm. The endogenous stages in cats fed sporulated oocysts were essentially similar, but merogony occurred 12–48 hours later than in the mouse-induced cycle.

Gametogony. Nery-Guimaraes and Lage (1973) said that the macrogametes averaged 15 x 13 μm and the microgamonts 13 x 9 μm; the latter contained more than 50 microgametes each.

Dubey (1979) found most of the macrogametes and microgamonts in the ileum. The early ones were elongate. He found gamonts of the mouse-induced cycle at 72–96 hours after inoculation. Mature microgamonts were 9–15 x 6–9 (mean, 11 x 8) μm in sections and up to 21.5 x 14 μm in smears; they contained up to 70 microgametes. Macrogametes were 11–18 x 5–13 (mean, 13 x 9) μm in sections and 18 x 16 μm in smears. The endogenous stages in cats fed sporulated oocysts were essentially similar, but gametogony occurred 12–14 hours later than in the mouse-induced cycle.

Prepatent Period. 4–7 days (Dubey and Frenkel, 1972; Frenkel and Dubey, 1972; Dubey and Streitel, 1976; Dubey, 1976); 4–5 days (Dubey, 1979).

Patent Period. Oocysts passed intermittently for at least 66 days (Dubey, 1979).

Pathogenicity. I. rivolta is only slightly if at all pathogenic in cats. Dubey and Streitel (1976) did not consider it to be a serious feline pathogen. Dubey (1979) found that newborn kittens fed 100,000 sporocysts or infected mice usually developed diarrhea in 3–4 days. He found desquamation of the tips of the villi and cryptitis of the ileum and cecum in them. Weaned kittens shed millions of oocysts but remained asymptomatic.

Immunity. Chessum (1972) found that no cross-immunity existed between *I. rivolta, I. felis,* and *Toxoplasma gondii.*

Cross-Transmission Studies. Patent infections do not occur in dogs fed *I. rivolta* oocysts from cats (Dubey, Miller, and Frenkel, 1970; Nery-Guimaraes and Lage, 1973; Guterbock and Levine, 1977) or in dogs fed occultly infected mice (Guterbock and Levine, 1977). Kotlan and Pospesch (1933) were unable to produce a patent infection in a young badger *Meles meles.* Nukerbaeva and Svanbaev (1974) were unable to produce patent infections in *Vulpes vulpes, Alopex lagopus,* or the mink with oocysts from the cat. However, as stated above, occult infections of various parenteral organs whose presence is revealed by feeding of tissues to cats occur in the house mouse, Norway (laboratory) rat, golden hamster, and domestic cat.

Dubey and Frenkel (1972) fed kittens sporocysts of *I. rivolta* and then killed them at intervals and fed suspensions of their extra-intestinal tissues to indicator kittens less than a day old. If the indicator kittens produced oocysts with a regular prepatent period, they concluded that coccidian stages were present in the extrain-testinal tissues of the donor cats. They found that the liver, spleen, and mesenteric lymph nodes were positive for *I. rivolta* for 21 days after inoculation, the brain and muscle for 12 days, and the lungs for 10 days. The prepatent period was 5–7 days whether the kittens consumed oocysts or extraintestinal tissues. They found coccidian stages either singly or in pairs, both intracellularly and free, in the mesenteric lymph nodes of 3 out of 10 kittens infected for 1–8 days.

Frenkel and Dubey (1972) found that *I. rivolta* produced non-patent infections in laboratory mice and that oocysts appeared in newborn kittens after the mice had been fed to them. The prepat-ent period between the feeding of the mice and the appearance of oocysts in the kittens was 5–6 days, the same as in kittens fed oocysts. The coccidia were present in the lungs, liver, spleen, and mesenteric lymph nodes of the mice for at least 67 days. The mero-zoites in the mesenteric lymph nodes were 10–13 x 5–8 μm; when ensheathed, they were 10–15 x 7–10 μm, and the sheath was 14–20 x 7–11 μm. The sporozoites the animals had ingested were much smaller—6–8 x 3–4 μm. Dubey and Streitel (1976) found that cats could be infected with *I. rivolta* by feeding them homogenized mice that had been fed oocysts 5 days before. The prepatent period in kittens fed mice was 4–8 (mean, 5.2) days; the time was 4–7 (mean 5.2) days for kittens fed oocysts.

Frenkel and Dubey (1972) also produced nonpatent infections in laboratory rats and golden hamsters. The organisms were in the lungs, livers, spleen, and/or lymph nodes for at least 10 days, and newborn kittens were infected by ingesting these organs.

There is no evidence that *I. rivolta* is transmitted transplacentally from mother cats to kittens (Dubey, 1977).

Cultivation. Fayer (1972) cultivated *I. rivolta* from cats in feline kidney and embryonic bovine kidney tissue culture cells, but did not succeed in cultivating it in Madin-Darby canine kidney cell

tissue cultures. He obtained 1 generation of multiplication (i.e.,
first-generation merozoites), but nothing beyond that.

Remarks. This seems to be the form that Yakimoff et al. (1933)
said that Triffitt (1927) found in the ocelot in the London zoo.
However, all she mentioned was the eyot-cat. Perhaps Yakimoff
et al. (1933) thought that this was an ocelot. However, "eyot" is
British dialect for islet, especially in a river such as the Thames,
and the term does not apply to the ocelot. Hence we are ignoring
this report in the belief that the eyot-cat was either *F. catus* or
perhaps *F. silvestris.*

Sarcocystis hirsuta Moulé, 1888

Synonyms. Miescheria cruzi Hasselmann, 1926 in part; *Sarcocystis fusiformis* Railliet, 1897 of Babudieri (1932) and *auctores*
in part; *Sarcocystis marcovi* Vershinin, 1975 in part; large form of
Isospora bigemina from cats of Gestrich, Mehlhorn and Heydorn,
1975; *Sarcocystis bovifelis* Heydorn et al., 1975; *Endorimospora
cruzi* (Hasselman, 1926) Tadros and Laarman, 1976.

Type Definitive Host. Domestic cat *Felis catus* (Heydorn and
Rommel, 1972a; Markus, Killick-Kendrick, and Garnham, 1974;
Rybaltovskii, Dudkina, and Rubina, 1973).

Other Definitive Host. European wild cat *F. silvestris.*

Intermediate Host. Ox *Bos taurus.*

Other Intermediate Hosts. None proven.

Location. Gamonts, gametes, zygotes, oocysts, and sporocysts
in lamina propria of villi of small intestine of cats. Location of
early meronts, if any, unknown. Sarcocysts (meronts) in skeletal
muscles of ox.

Geographic Distribution. Worldwide.

Prevalence. Common. Because of the similarity between the
sporocysts of *S. hirsuta, S. gigantea, S. porcifelis,* and *S. muris,*
and because of the confusion existing until very recently regarding
the differences between these 4 species and the so-called large form
of *I. bigemina,* no exact figures can be given on the prevalence of
S. hirsuta in cats. It is known that one can obtain sporocysts of
this species routinely by feeding bovine muscle to cats (e.g., Heydorn and Rommel, 1972), so *S. hirsuta* is obviously common. Us-

ing tryptic digestion, Boch, Laupheimer, and Erber (1978) found *Sarcocystis* spp. in 99.7% of 1,020 cattle from 11 slaughter houses in south Germany; using the trichinoscope they found it in only 57.9%. They found *S. hirsuta* in 34.5% of 817 positive cattle. In studies of cat feces, sporocysts of one or all 4 of the feline species of *Sarcocystis* have been found in 2% of 50 cats in Utrecht, Holland, by Nieschulz (1925), in 0.2% of 1,000 cats from a Columbus, Ohio, humane shelter by Christie, Dubey, and Pappas (1976) and in none of 217 cats in Illinois by Guterbock and Levine (1977).

Oocyst Structure. Oocysts in the lamina propria of the cat are 12–17 x 11–14 (mean, 15 x 12) μm in Giemsa-stained smears (Heydorn and Rommel, 1972). They have a thin, smooth, colorless, presumably 1-layered wall and apparently lack a micropyle, residuum, or polar granule. They sporulate in the host, and the oocyst wall is stretched between the 2 sporocysts, producing a dumbbell-like appearance. The sporocysts have a smooth wall thicker than that of the oocyst, and are 11–14 x 7–9 (mean, 12.5 x 8) μm (Heydorn and Rommel, 1972; Rommel et al., 1974; Gestrich, Heydorn, and Baysu, 1975). They contain 4 sausage-shaped sporozoites and a residuum when passed by the cat. They have no Stieda body.

Sporulation. The process has been described above. Sporulation takes place in the cat, sporulated sporocysts being passed in the feces.

Merogony. It is not known how many meront generations there are in the endothelial cells of cattle, as is known for *S. cruzi.* Later on, meronts (sarcocysts) can be found in the skeletal muscles of cattle. Gestrich, Mehlhorn, and Heydorn (1975) and Mehlhorn et al. (1975) described the sarcocysts that were formed in calves fed sporozoites from cats that had been fed raw esophageal muscle from cattle. The sarcocysts in the calves 98 days after being fed sporozoites were always within a muscle fiber and never surrounded by fibrillar layers of host origin. They were limited by a single unit membrane thickened at many places by osmiophilic material; they called this complex a primary wall. It was up to 24 nm thick and folded regularly to form palisadelike protrusions about 4.7 μm long and 1.5 μm in diameter. These protrusions contained about 200–300 parallel fibrils running from the tip of each protrusion into the interior of the sarcocyst; the fibrils were about 15–18 nm in diameter and seemed to be united in several cases to form larger tubules about 50–60 nm in diameter.

At this time there were both metrocytes and merozoites in chamberlike hollows of the ground substance of the sarcocysts. The metrocytes formed merozoites, presumably by endodyogeny and possibly by binary fission as in *S. gigantea* (see below), so that by 120 days only merozoites were present. The sarcocyst wall was up to 5.4 μm thick and appeared radially striated by light microscopy (Heydorn et al., 1975).

The metrocytes were globular, about 12–14 x 5–7 μm, with a typical 3-layered pellicle, deep micropores, a conoid, polar ring with 22 anchored subpellicular microtubules, very few rhoptries and micronemes, a Golgi complex anterior to the large nucleus, a spherical nucleolus consisting of granular and fibrillar zones, and chromosomal structures in 2 different stages: a few large, dense plaques of the condensed stage, and small, dense granules, 30–40 nm in diameter, arranged spherically within the karyoplasm as the extended stage. They found no significant difference except in size between the fine structure of the metrocytes in the sarcocysts of *S. hominis, S. cruzi,* and *S. hirsuta.*

The merozoites were about 13–17 x 2.5–3 μm and had 22 subpellicular microtubules, normally 1 micropore at the apical pole evidently "ingesting small vesicles by large vacuoles," several rhoptries (12 in single longitudinal sections), and numerous micronemes filling the anterior 1/3 of the cell. They also had all the other organelles of other coccidia.

Gametogony. Sheffield and Fayer (1978) found that the oocysts develop extracellularly in the lamina propria of the cat small intestine and have no wall-forming bodies. The process has not been described in detail. Heydorn and Rommel (1972) first found developmental stages in the cat 6 hours after feeding microscopic sarcocysts from the esophagus of cattle. They were rounded and in the lamina propria under the epithelium of the villi of the whole small intestine. After 3 days live macrogametes there were 11–14 x 8–9 (mean, 12.5 x 8) μm. After 5 days live oocysts in the lamina propria were 12–17 x 11–14 (mean, 15 x 12) μm. Heydorn and Rommel (1972) saw no microgamonts or meronts.

Prepatent Period. 7 days (Markus, Killick-Kendrick, and Garnham, 1974); 7–9 days (Rommel et al., 1974; Gestrich, Heydorn, and Baysu, 1975); 8–9 days (Rybaltovskii, Dudkina, and Rubina, 1973); or 10–30 days (Dubey and Streitel, 1976).

Patent Period. 6–17 days (Dubey and Streitel, 1976), more than 6 weeks (Heydorn and Rommel, 1972; Rommel et al., 1974; Gestrich, Heydorn, and Baysu, 1975) or 60 days (Suteu and Coman, 1973).

Pathogenicity. This species is not or only slightly pathogenic for calves. Calves fed 2 million sporocysts survived and had little or no clinical reaction (Gestrich, Heydorn, and Baysu, 1975). Panasyuk et al. (1971) and Sominskii, Panasyuk, and Vilkova (1971) said that this species was highly pathogenic for chickens, but Rybaltovskii, Dudkina, and Rubina (1973) and Lane and Levine (unpublished) were unable to confirm this contention.

Immunity. Little if anything is known about the development of immunity against *S. hirsuta,* either in the ox or cat.

Cross-Transmission Studies. So far as is known, only cats pass sporocysts (or oocysts) after oral inoculation. The following have been found not to do so: laboratory mouse and guinea pig (Suteu and Coman, 1973), laboratory rat (Aryeetey and Piekarski, 1976), rhesus monkey *Macaca mulatta* and baboon *Papio cynocephalus* (Heydorn, Gestrich, and Janitschke, 1976), splenectomized chimpanzee *Pan troglodytes* (Markus, Killick-Kendrick, and Garnham, 1974), chicken (Suteu and Coman, 1973; Rybaltovskii, Dudkina, and Rubina, 1973; Lane and Levine [unpublished]).

Remarks. Gestrich (1974) found that the sarcocysts of *S. hirsuta* in bovine diaphragm were killed by storage for 3 days at –20 C or by heating to 65–70 C for 10 minutes. Gestrich and Heydorn (1974) expanded on this finding, saying that the sarcocysts were still infectious for cats after 18 days at 2 C but not after 3 days at –20 C; in steaks they were killed only if the meat had been heated to an internal temperature of 65–70 C. They were still infectious in medium-done steaks.

Sarcocystis gigantea (Railliet, 1886) Ashford, 1977

Synonyms. Balbiania gigantea Railliet, 1886; *Sarcocystis ovifelis* Heydorn et al., 1975; *Sarcocystis tenella* (Railliet, 1886) Moulé,

1886 in part; *Endorimospora tenella* (Railliet, 1886) Tadros and Laarman, 1976 in part.

Type Definitive Host. Domestic cat *Felis catus.*

Other Definitive Host. Red fox *Vulpes vulpes* (experimental), possibly coyote *Canis latrans.*

Intermediate Host. Domestic sheep *Ovis aries.*

Other Intermediate Host. Rocky mountain bighorn sheep *O. canadensis.*

Location. Gamonts, gametes, zygotes, oocysts, and sporocysts in lamina propria of villi of small intestine of cat. Location of early meronts, if any, unknown. Sarcocysts (meronts) in skeletal muscles of sheep, especially in wall of esophagus.

Geographic Distribution. Worldwide.

Prevalence. This species is extremely common, having been reported from 50%–100% of the sheep examined (Scott, 1943; Destombes, 1957; Grassé, 1953). It has also been reported from domestic goats, in which it is uncommon (Reichenow, 1953), and whether goats have this species is unknown. Dubey, Fayer, and Seesee (1978) found *"Sarcocystis"* sp. sporocysts (some of which may have been *S. gigantea*) in 53% of 169 coyotes in Montana.

Oocyst Structure. Mehlhorn (1974) and Mehlhorn and Scholtyseck (1974) studied the oocysts and sporocysts of *S. gigantea* in the cat intestinal wall. They treated one cat with the corticosteroid Prednisone before inoculation with about 100 sarcocysts from the sheep and killed the cat 9 days after inoculation. The intestine was not inflamed but contained a considerable number of oocysts, the great majority in the posterior small intestine, in large parasitophorous vacuoles just beneath the epithelial cells of the villi. They had no micropyle and were about 15–18 x 10–14 μm and contained sporocysts 10–12 x 6–8 (mean, 11 x 7) μm. Most of the oocysts were sporulated, but the few unsporulated ones had a smooth, colorless, 1-layered wall about 0.25 μm thick; it shrank to 0.1 μm during sporulation. The oocysts apparently have no micropyle, residuum, or polar granule. In the feces of untreated cats the sporocysts were 10–14 x 8–10 (mean, 12 x 9) μm; they had no Stieda body but did have a residuum. The sporocysts were said to be 11–14 x 8–9 (mean, 12 x 8) μm by Rommel, Heydorn, and Gruber (1972). This mean was confirmed by Rommel et al. (1974) and

Munday and Rickard (1974). They have a smooth, colorless wall, no Stieda body, and contain a residuum and 4 sausage-shaped sporozoites with a clear globule at one end. Rommel, Heydorn, and Gruber (1972) occasionally saw a pair of sporocysts held together by the thin oocyst wall.

Sporulation. As already stated, this occurs in the cat; Mehlhorn (1974) and Mehlhorn and Scholtyseck (1974) described it.

Merogony. It is not known whether there is an early generation of meronts in the visceral or other organs of sheep. The sarcocysts in the sheep muscle are relatively ellipsoidal and up to 1 cm long. They grow slowly, and few are found in sheep under 3 years of age in Australia (Munday and Rickard, 1974). Their fine structure and that of the bradyzoites within them was described in detail by Sénaud (1967), Mehlhorn and Scholtyseck (1973), Porchet-Henneré and Ponchel (1974), and Mehlhorn, Hartley, and Heydorn (1976), among others. The sarcocyst wall is composed of 2 distinct layers, a primary and a secondary wall (Mehlhorn and Scholtyseck, 1973). The secondary wall, formed by the host cell, is composed of 2 layers; the outer is lamellar cytoplasm and within it is a layer of muscular tissue containing many nuclei and mitochondria in young sarcocysts. In larger sarcocysts the myofibrils are almost completely disintegrated.

The primary wall limits the sarcocyst proper and is formed by the parasite. It is osmiophilic, about 25 nm wide, and has many villuslike folds and vesiclelike invaginations into the sarcocyst interior. The folds form cauliflowerlike protrusions, which contain microtubules. Beneath the primary sarcocyst wall is a zone of fine, granular ground substance that extends as septa into the interior of the sarcocyst, dividing it into compartments.

In the younger sarcocysts the periphery is occupied by metrocytes, which are globular cells about 15–20 μm long, with a deeply invaginated pellicle; they contain several micropores. These cells have a typical conoid, polar ring with anchored microtubules and Golgi complex. The nucleus has a nucleolus and a globular accumulation of electron-dense granules that are apparently not present in merozoites. Mehlhorn and Scholtyseck (1973) did not see rhoptries or micronemes in them, but others apparently have (see Scholtyseck, 1973). This last author said that "their major features are a deeply folded cell surface, a widely distributed system of vesicles and lacunae in the cytoplasm and a nucleolus surrounded by a spiral structure, which resembles that of a chromosome.

Sénaud (1967) recognized 2 types of metrocyte—eumetrocytes reproducing by fission ("scissiparity"), and heterometrocytes form- ing 2 endodyocytes by endodyogeny; however, others do not rec- ognize these types and say that reproduction is by endodyogeny. At any rate, the metrocytes divide repeatedly, each time becoming more like banana-shaped merozoites, and are finally recognized as such. The merozoites (bradyzoites) occur in older sarcocysts and eventually replace the metrocytes entirely. They are about 12–15 x 3–4 μm, with 22 subpellicular microtubules, 11 riblike elements (consisting of rows of granules) on the outer surface, a conoid, up to 11 rhoptries, and about 400 micronemes (Müller, Mehlhorn, and Scholtyseck, 1973; Scholtyseck, Mehlhorn, and Müller, 1973). They have a single branched mitochondrion. Mehlhorn et al. (1975) found that the micronemes appear in small vacuoles at the edges of 2 large granular vacuoles in each daughter cell during endodyog- eny. Later these vacuoles are divided into numerous vesicular spiral formation centers, which produce micronemes at the poles of the young merozoites. Rhoptries originate from what they called densifications within the same large vacuoles that give rise to the micronemes. Thus, the rhoptries and micronemes seem to originate from the same vacuolar substance, even though they differentiate into different structures. Porchet-Henneré (1975) said that the micronemes look like rice grains and, contrary to previous state- ments, the grains are independent of one another. She also said that there are two additional microtubules inside the conoid, which end abruptly a little behind the conoid, 2 (perhaps 3) polar rings, and that the conoid itself consists of about 20 oblique, spirally coiled microtubules.

Acid phosphatase is present in the endoplasmic reticulum of the metrocytes and merozoites (Mehlhorn and Scholtyseck, 1973a), alkaline phosphatase along the outer membrane, and adenosine triphosphatase in the endoplasmic reticulum, in the perinuclear space, between the 2 inner membranes of the 3-layered pellicle, and only slightly in the mitochondrion (Mehlhorn and Scholtyseck, 1974b).

Gametogony. This process was seen, but not described separately, in the cat small intestine by Becker, Mehlhorn, and Heydorn (1979). It occurs in the lamina propria of the villi and takes less than a day.

Prepatent Period. 11-12 days (Rommel, Heydorn, and Gruber, 1972); 10-11 days (Mehlhorn and Scholtyseck, 1974b); 11-14 days (Rommel et al., 1974).

Patent Period. 4-53 days (Rommel, Heydorn, and Gruber, 1972); up to 14 days (Mehlhorn and Scholtyseck, 1974a); more than 6 weeks (Rommel et al., 1974).

Pathogenicity. This species is apparently only slightly, if at all, pathogenic for either lambs or cats.

Immunity. Immunity is not produced in cats by previous infection. Rommel, Heydorn, and Gruber (1972) found that a second infection of cats was followed by a prepatent period of 11-15 days and a patent period of 25-47 days; the times were 11-12 and 4-53 days, respectively, in cats upon primary infection.

Aryeetey and Piekarski (1976) found that freezing or cooking *S. gigantea* sarcocysts destroyed their ability to induce the formation of antibodies in rats fed frozen or cooked mutton.

Cross-Transmission Studies. Ashford (1977) transmitted this species to the red fox *V. vulpes;* its sporocysts were 13-14 x 9-10 μm. The dog (Rommel, Heydorn, and Gruber, 1972) and laboratory rat (Aryeetey and Piekarski, 1976) cannot be infected.

Cultivation. Dubremetz, Porchet-Henneré, and Parenty (1975) cultivated *S. gigantea* in embryonic sheep kidney tissue culture, inoculating it with sarcocysts from the sheep esophagus. The crescent-shaped "stage 1" bradyzoites entered the cells and transformed into thicker, oblong forms (called "stage 2") and then into ovoid "stage 3" parasites.

Sarcocystis porcifelis Dubey, 1976

Synonym. Sarcocystis miescheriana (Kühn, 1865). Labbé, 1899 in part.

Type Definitive Host. Domestic cat *Felis catus.*

Intermediate Host. Domestic pig *Sus scrofa.*

Location. Gamonts, gametes, zygotes, oocysts, and sporocysts presumably in lamina propria of villi of small intestine of cat. Location of early meronts, if any, unknown. Sarcocysts (meronts) in skeletal muscles of pigs.

Geographic Distribution. So far this species has been recognized only in the USSR (Golubkov, Rybaltovskii, and Kislyakova, 1974), but it probably occurs throughout the world.

Prevalence. Unknown.

Oocyst Structure. Apparently only sporocysts are passed in the feces of cats. They average 13.5 x 8 μm and contain 4 sporozoites, averaging 9.5 x 4 μm, plus a residuum (Golubkov, Rybaltovskii, and Kislyakova, 1974).

Sporulation. As in other species of *Sarcocystis,* sporulation occurs in the body of the definitive host.

Merogony. Not followed. Sarcocysts occur in pig muscles.

Gametogony. Not followed.

Prepatent Period. 4–9 days (Golubkov, Rybaltovskii, and Kislyakova, 1974).

Patent Period. 10–15 days (Golubkov, Rybaltovskii, and Kislyakova, 1974).

Pathogenicity. This species is apparently not pathogenic for cats. However, it is for swine, causing poor growth, diarrhea, myositis, and lameness (Golubkov, Rybaltovskii, and Kislyakova, 1974).

Immunity. Unknown.

Remarks. This species was named by Dubey (1976) on the basis of the Russian findings mentioned above.

Sarcocystis fusiformis (Railliet, 1897) Bernard and Bauche, 1912

Synonyms. Balbiania fusiformis Railliet, 1897; *Sarcocystis blanchardi* Doflein, 1901; *Sarcocystis siamensis* von Linstow, 1903; *Sarcocystis bubali* Willey, Chalmers, and Philip, 1904.

Type Definitive Host. Domestic cat *Felis catus.*

Other Definitive Hosts. Unknown.

Intermediate Host. Water buffalo *Bubalus bubalis.*

Other Intermediate Hosts. None.

Location. Gamonts, gametes, zygotes, oocysts, and sporocysts in cat small intestine wall. Sarcocysts in water buffalo muscles.

Geographic Distribution. Africa (Egypt), Asia (Malaysia).

Prevalence. Apparently common.

Oocyst Structure. Oocysts 17–18 x 13 μm (Hilali and Scholty-seck, 1978). Sporocysts in cat feces 12–14 x 7–9 (mean, 13 x 8) μm (Dissanaike et al., 1977), 12–13 x 8–10 μm (Hilali and Scholty-seck, 1978).

Sporulation. Occurs in the definitive host.

Merogony. Railliet (1897) named this species from sarcocysts collected from the esophageal muscles of the water buffalo in Egypt. They differed notably in external appearance from those of *S. tenella.* They were separated very easily from the surrounding muscle tissue, were whitish, fusiform bodies with one end more pointed than the other, and attained a length of 5–15 mm and a width of 2–4 mm. He named them *Balbiania fusiformis* and re-marked that they were probably the same as those seen in buffalo in Java by J. de Jongh. The sarcocysts have cauliflowertype cyto-phaneres (Dissanaike et al., 1977).

Ghaffar, Hilali, and Scholtyseck (1978) found large and small sarcocysts (apparently both of *S. fusiformis*) in the esophageal muscles of water buffalos from Egypt. The large ones were 7–30 x 3–7 mm and the small ones, 1.3–5 x 0.7–2 mm. Both were com-partmented. The metrocytes of both types were mostly at the peri-phery and multiplied by endodyogeny; they were 8–9 x 5 μm and most had deep invaginations in their pellicle. The merozoites of both types were 15–17 μm long and had a conoid, 8–13 rhoptries, about 300 micronemes, 22 subpellicular microtubules, 1 micropore, a thick-walled oval vesicle, a mitochondrion, many amylopectin granules, large lipid droplets, a nucleus at the beginning of the posterior 1/3 of the body, ER, and ribosomes. The walls of both types of sarcocyst were similar. They were irregularly folded and contained numerous cauliflowerlike projections or protrusions up to 3.4 μm long and 0.87 μm wide, and also had small, droplike invaginations. The protrusions contained many fibrillar elements.

Gametogony. Scholtyseck and Hilali (1978) studied the fine structure of the sexual stages in the small intestine of the cat. They lie in a parasitophorous vacuole in the epithelial cells or in the cells beneath the epithelium. At 13–14 days the intracellular merozoites are about 15 μm long. They contain more than 13 osmiophilic bodies resembling rhoptries both at the anterior and posterior ends. They turn into macrogametes or microgamonts. The macrogametes

contain many amylopectin granules plus wall-forming bodies I and
II.
Prepatent Period. 17–33 days (Dissanaike, in lit.) or 10–13 days
(Hilali and Scholtyseck, 1978).
Patent Period. Unknown.
Pathogenicity. Unknown.
Immunity. Unknown.
Cross-Transmission Studies. Dissanaike (in lit.) failed to infect 4
monkeys with sarcocysts from the water buffalo.

Sarcocystis muris (Blanchard, 1885) Labbé, 1899

Synonyms. Miescheria muris Blanchard, 1885; *Coccidium bigeminum* var. *cati* Railliet and Lucet, 1891; *Lucetina cati* (Railliet and
Lucet, 1891) Henry and Leblois, 1926; *Sarcocystis musculi* Blanchard, 1885 of Kalyakhin and Zasukhin, 1975 *lapsus calami; Endorimospora muris* (Blanchard, 1885) Tadros and Laarman, 1976;
probably *Isospora cati* (Railliet and Lucet, 1891); possibly *Cryptosporodium [sic]* sp. Dubey and Pande, 1963 from *Felis chaus.*
Type Definitive Host. Domestic cat *Felis catus.*
Other Definitive Hosts. Possibly Indian jungle cat *F. chaus.* Dubey
and Pande (1963a) found cysts that they called *Cryptosporodium
[sic]* sp. in the feces of this animal in India. They were ellipsoidal,
10–12 x 7–8 (mean, 11 x 7) μm, with a 2-layered wall, outer layer
thick and pale yellow, inner layer, green. They contained 4 naked
sporozoites, 7–9 x 1–2 μm, with a clear globule at the broad end.
The cysts were illustrated without a Stieda body but with a residuum. They were apparently already sporulated when passed in
the feces. This was undoubtedly a species of *Sarcocystis* rather
than *Cryptosporidium,* and possibly *S. muris.*
    Perhaps this species also infects the bobcat *Lynx rufus.* Duszynski
and Speer (1976) described excystation of what they called the
large form of *Isospora bigemina* from the bobcat, but they gave no
measurements. The cysts that they studied were 229 days old when
they exposed them to TST (0.25% trypsin-0.75% sodium taurocholate solution). After 23 hours at 37 C, 23% of 60 free sporocysts had excysted, but they found no sporozoites.

Intermediate Host. House mouse *Mus musculus.*

Other Intermediate Hosts. This species has also been reported from the Norway rat *Rattus norvegicus*, black rat *R. rattus*, voles *Microtus arvalis* and *M. agrestis*, red-backed mouse *Clethrionomys glareolus*, and rice rat *Oryzomys goeldi.* However, it is unlikely that these animals are hosts of *S. muris*, since Ruiz and Frenkel (1976) were unable to infect the Norway rat, golden hamster, or guinea pig with *S. muris.*

Location. Gamonts, gametes, zygotes, oocysts, and sporocysts at level of basement membrane and in lamina propria of intestine of the cat. First-generation meronts and merozoites (tachyzoites) in liver of house mouse. Second-generation meronts (sarcocysts) containing metrocytes and merozoites (bradyzoites) in striated muscles of house mouse but not in myocardium (Ruiz and Frenkel, 1976).

Geographic Distribution. Presumably worldwide.

Prevalence. Common. Surveys of cats for this species cannot be relied upon because, so far as is known at present, the sporocysts are similar to those of other species of *Sarcocystis* in cats. *S. muris* often occurs in mice. Šebek (1975), for instance, found it in 5% of 383 wild *M. musculus* in Czechoslovakia.

Oocyst Structure. The oocysts have apparently not been described in detail. Ruiz and Frenkel (1976) found them in infected cats and said that they persisted in the lamina propria for an indefinite number of days. Beginning 5–6 days after inoculation, they contained 2 sporulated sporocysts. They have a thin wall that is stretched between the sporocysts, causing the latter to lie in pairs and to have a dumbbell shape. They are 9–12 x 7–9 (mean, 10 x 8.5) μm, have smooth, colorless walls without a Stieda body, and contain a few residual granules and 4 slightly curved sporozoites about 10 x 2 μm with a central nucleus.

The fine structure of the sporozoites is similar to that of *Toxoplasma gondii* described below (Sheffield and Melton, 1974). They have relatively large mitochondria, which sometimes occur anterior to the nucleus, have rare amylopectin granules, and their posterior region is filled with moderately dense spherical granules about 40 nm in diameter that are similar to those of other coccidia. These granules are presumably storage products, since they disappear after the sporozoite has penetrated a host cell.

Sporulation. Occurs in the cat intestinal wall beginning 5–6 days after inoculation. The sporocysts break out of the oocysts and are passed in the feces.

Merogony. According to Ruiz and Frenkel (1976), the first-generation meronts occur in the liver of house mice 11–17 days after the feeding of sporocysts from the cat. These meronts contain tachyzoites with a central nucleus; they did not give their number, but in a photomicrograph of a liver section they published there was a wheel of 28 tiny tachyzoites with what appeared to be a residuum, so the true number of tachyzoites per meront must have been much greater. Ruiz and Frenkel (1976) first saw sarcocysts in the skeletal muscle 28 days after inoculation; they were 25–40 μm long in sections and consisted of a few metrocytes. By 6 weeks some of the sarcocysts were 400 μm long and contained hundreds of organisms. They continued to grow; when they were 5–6 or more millimeters long, they contained thousands of bradyzoites 14–16 x 4–6 μm in compartments. The bradyzoites were infectious for cats beginning at about 76 days. The sarcocysts continued to grow, becoming more than 1 cm long and replacing the muscle fibers.

The sarcocyst wall is smooth, without radial spines (cytophaneres) (Ruiz and Frenkel, 1976). Its fine structure was studied by Mehlhorn, Hartley, and Heydorn (1976), Sheffield, Frenkel, and Ruiz (1977), and Fedoseenko and Levit (1979). Mehlhorn, Hartley, and Heydorn (1976) said that the primary wall is up to 60 nm thick and has very many short, narrow protrusions up to 0.15 μm long and that are not visible under the light microscope. The zone of ground substance is up to 2 μm thick. Septa divide the sarcocyst into compartments. There is no secondary wall around the sarcocysts. According to Fedeseenko and Levit (1979), the sarcocysts are contained in a parasitophorous vacuole. Closely spaced spherical blebs are formed from this membrane and extend into the muscle cell cytoplasm. The cavity of the bleb is filled with a dense substance. The same substance occupies the vacuolar space immediately adjacent to the membrane, while the rest of the vacuole is filled with a moderately dense matrix within which the parasites develop. Forty days after inoculation, only metrocytes are present; they can be recognized by their ovoid shape, lightly stained cytoplasm, amylopectinlike granules, and lack of micronemes. They divide by a process resembling endodyogeny, and eventually produce bradyzoites. Fedoseenko and Levit (1979) said that the metrocytes give rise to 4 daughter cells simultaneously. By 78 days after inoculation the sarcocyst is infective for cats. At this time the few remaining metrocytes are at the periphery of the sarcocyst, but most of the cells have elongated and are bradyzoites containing micronemes, dense granules, amylopectin, and a few rhoptries. There is also a conoid. Many smooth endoplasmic reticulum vesicles accumulate in the muscle cell cytoplasm at the periphery of the enlarging sarcocyst, but there was no significant destruction of muscle fibers in specimens fixed 40-325 days after inoculation. Sheffield, Frenkel, and Ruiz (1977) saw unusual lamellar structures in some parasitized muscle cells and intracystic tubules in some sarcocysts. No merogony occurs in the definitive host (Becker, Mehlhorn, and Heydorn, 1979).

Gametogony. According to Ruiz and Frenkel (1976), elongate organisms begin to penetrate the intestinal epithelium of the cat, usually through the goblet cells, 3-6 hours after ingestion of sarcocysts. They had rounded up by 9 hours, and macrogametes and a few microgamonts 6.5 μm in diameter were seen by 12 hours. At 24 hours macrogametes, which were possibly fertilized, had moved to the level of the basement membrane or on into the lamina propria. Young oocysts were present at 2 days; they contained 1 or 2 sporoblasts at 3 days and sporocysts at 4 days. Sporulated sporocysts were present at 5-6 days. The oocysts ruptured, releasing the sporocysts, but some intact oocysts remained in the lamina propria for an indefinite number of days.

Prepatent Period. 7-26 days, but usually 7-10 days (Ruiz and Frenkel, 1976).

Patent Period. 3-81 days, but usually 5-10 days (Ruiz and Frenkel, 1976).

Pathogenicity. S. muris is apparently not pathogenic for the cat, but is is for the mouse. According to Ruiz and Frenkel (1976), mice with heavy infections moved with difficulty. Myositis and sometimes muscle necrosis, apparently due to degeneration or rupture of the sarcocysts, were present. In heavy infections the sarcocysts outlined the major muscle groups. They were easily visible to the naked eye.

Immunity. Previous infection does not immunize either cat or mice, since they can be infected repeatedly (Ruiz and Frenkel, 1976).

The IFA (indirect fluorescent antibody) test was used by Ruiz and Frenkel (1976) for serologic studies. *S. muris* bradyzoites were used for the antigen. Titers of 1:512–1:4,000 were produced in both mice and cats as the result of infection. There was little if any cross-reaction with *Toxoplasma gondii* or *Besnoitia* sp.

Cross-Transmission Studies. Ruiz and Frenkel (1976) were unable to infect the laboratory rat *Rattus norvegicus,* golden hamster *Mesocricetus auratus,* or guinea pig *Cavia porcellus* with *S. muris,* giving credence to the idea that these animals have different species of *Sarcocystis.*

Cultivation. Mehlhorn, Becker, and Heydorn (1978) and Becker, Mehlhorn, and Heydorn (1979) cultivated *S. muris* from bradyzoites to oocysts and sporocysts in cat lung and very poorly in dog kidney cell cultures but not in human fibroblast or pig kidney cultures.

Remarks. This was the first species of *Sarcocystis* to be found. Miescher discovered its sarcocysts in the muscles of house mice in 1842.

Ruiz and Frenkel (1976) found that this species, like all known species of *Sarcocystis,* has an obligatory 2-host life cycle. It cannot be transmitted from mouse to mouse or from cat to cat.

Sarcocystis leporum Crawley, 1914

Type Definitive Host. Domestic cat *Felis catus* (experimental).
Other Definitive Host. Raccoon *Procyon lotor* (experimental).
Intermediate Host. Cottontail *Sylvilagus floridanus.*
Other Intermediate Hosts. Cottontails *S. nuttalli* and *S. palustris.*
Location. Presumably intestine of cat. Muscles of cottontails.
Geographic Distribution. North America (United States).
Prevalence. Common in cottontails. Fayer and Kradel (1977) found sarcocysts in 15 of 18 *S. floridanus* from Pennsylvania; 11 of these had only microscopic sarcocysts.

Oocyst Structure. Sporulated sporocysts and occasionally oocysts are shed in the feces of cat. The sporocysts found by Fayer and Kradel (1977) in cat feces were ellipsoidal, 13–17 x 9–11 (mean, 14 x 9) μm, without a Stieda body, with a large residuum. The sporozoites were elongate, with one end broad and the other tapered, apparently without a clear globule. The sporocysts found by Crum and Prestwood (1977) in the cat were ellipsoidal, 13 x 9-11 (mean, 13 x 10) μm, thin-walled, without a Stieda body, with a residuum. The sporocysts found by Crum and Prestwood (1977) in a raccoon were 11-14 x 9-11 (mean, 13 x 9) μm.

Sporulation. In the host.

Merogony. In cottontail muscles. Sarcocysts (meronts) were both macroscopic and microscopic. The macroscopic sarcocysts seen by Crum and Prestwood (1977) were 243 μm in diameter.

Gametogony. Unknown.

Prepatent Period. 10-25 days in cat (Fayer and Kradel, 1977); 56 days in cat (Crum and Prestwood, 1977).

Pathogenicity. Presumably none for either cottontail or cat.

Immunity. Fayer and Kradel (1977) were able to reinfect a cat.

Cross-Transmission Studies. Fayer and Kradel (1977) were unable to infect dogs with sarcocysts in cottontail muscle or to infect domestic rabbits *Oryctolagus cuniculus* with 200-75,000 sporocysts from the cat.

Sarcocystis cuniculi Brumpt, 1913

Type Definitive Host. Domestic cat *Felis catus* (experimental).

Intermediate Host. European rabbit *Oryctolagus cuniculus.*

Location. Presumably intestine of cat. Muscles of rabbit.

Geographic Distribution. Europe.

Prevalence. Presumably common.

Oocyst Structure. Sporulated sporocysts and oocysts are shed in the feces. Sporocysts 13 x 10 μm, with residuum. Sporozoites sausage-shaped (Tadros and Laarman, 1977).

Sporulation. In the definitive host.

Merogony. In rabbit muscles. Sarcocysts (meronts) are up to several millimeters long and 0.5 mm wide, compartmented, with a wall containing numerous fine, tall projections up to 11 μm high, tightly packed together into a luxuriant pile; these projections are seen with the electron microscope to be fingerlike and to contain numerous fibrils (Tadros and Laarman, 1977). The sarcocysts contain rounded metrocytes 4-5 μm in diameter and banana-shaped bradyzoites 11-12 μm long, resembling those of *S. gigantea* (Tadros and Laarman, 1977).

Prepatent Period. 9 days (Tadros and Laarman, 1977).

Cross-Transmission Studies. Tadros and Laarman (1977) were unable to infect the dog, a hand-reared fox *Vulpes vulpes*, a weasel *Mustela nivalis*, or a kestrel *Falco tinnunculus* by feeding sarcocysts from rabbits.

Sarcocystis cymruensis Ashford, 1978

Type Definitive Host. Domestic cat *Felis catus.*

Type Intermediate Host. Wild Norway rat *Rattus norvegicus.*

Other Intermediate Host. Possibly house mouse *Mus musculus;* Atkinson (1978) infected both mice and rats, but Ashford (1978) was unable to infect mice.

Location. Lamina propria of small intestine of cat; skeletal muscles of rat, but not in heart or tongue.

Geographic Distribution. Europe (England, Wales).

Prevalence. Atkinson (1978) found this species in 1 of 5 wild *R. norvegicus* in Wales.

Oocyst Structure. According to Ashford (1978), sporulated sporocysts are passed in cat feces. They are like those of *S. bigemina,* ellipsoidal, with 4 sporozoites and a residuum, without a Stieda body, averaging 10.5 x 7.9 μm. Atkinson (1978) said that the oocysts were 15 x 11 μm, the sporocysts 11 x 8 μm, and the sporozoites in fixed material 12 x 6 μm.

Sporulation. In the cat intestine.

Merogony. Atkinson (1978) found meronts 25 x 18 μm in the lungs and liver of a rat killed 12 days after feeding. He found free zoites 6 x 2 μm in the blood of a rat 11–15 days after feeding oocysts. The rat also had ovoid zoites 5 x 3 μm in its leukocytes at this time. Ashford (1978) said that 3 months after feeding, the sarcocysts in the rat muscles were about 0.5 mm long, not visible to the naked eye, and contained mostly metrocytes. At 6 months they were 3–13 (mean, 6.5) mm long and contained a few metrocytes restricted to the periphery of the sarcocyst, plus merozoites. At 9 months the sarcocysts were up to 5 cm long. They were thin-walled, not clearly septate, and were packed with merozoites 10 x 3.5 μm in fixed, stained sections. The merozoites had a double outer membrane, 4 rhoptries, micronemes, conoid, 23 subpellicular microtubules, about 15 large osmiophilic vacuoles, a variable number of small, clear, or intermediate-sized vacuoles, a micropore, and a large, central nucleus. Atkinson (1978) said that the sarcocysts were up to 3 cm long and had a wall 1 µm thick, with many small invaginations on the outer surface.

Gametogony. According to Ashford (1978) and Atkinson (1978), sexual stages are produced in the cat intestine.

Prepatent Period. 4 days or more (Ashford, 1978).

Patent Period. Up to 100 days (Ashford, 1978).

Immunity. Ashford (1978) infected a cat 3 times, but sporocyst production was reduced in the later infections.

Cross-Transmission Studies. Ashford (1978) was unable to infect the dog or the ferret *Mustela putorius* with sarcocysts from the rat.

Remarks. Ashford (1978) found that this species produced relatively few sporocysts in the cat and thought that perhaps another carnivore was the normal definitive host.

Sarcocystis sp. Eisenstein and Innes, 1956

Definitive Host. Unknown.
Intermediate Host. Domestic cat *Felis catus.*
Location. Striated muscles.
Geographic Distribution. Europe (Germany), North America.
Merogony. Only sarcocysts known.

Sarcocystis sp. Munday et al. 1978

Definitive Host. Feral domestic cat *Felis catus.*
Intermediate Host. Unknown.
Location. Feces.
Geographic Distribution. Australia.

Prevalence. Munday et al. (1978) found this form in 1 of 55 feral cats in Australia.

Oocyst Structure. Munday et al. (1978) mentioned no oocysts but said that they saw sporocysts 13–14 x 8.5 µm in the cat's feces.

Sporulation. In the host's intestine.

Remarks. Munday et al. (1978) said it was possible that this was *Frenkelia.*

Sarcocystis sp. Janitschke, Protz, and Werner, 1976

See the discussion of this species under Dog *Canis familiaris.*

Sarcocystis spp. Golubkov, 1979

Golubkov (1979) said that he found sarcocysts of *Sarcocystis horvathi* in the domestic chicken and *S. rileyi* in the domestic duck in the USSR and produced sporocysts in dogs and cats by feeding them sarcocysts from both species of birds. The prepatent period (apparently for both) was 10 days in the dog and 11 days in the cat. The patent period (apparently for both) was 21–23 days in the dog and 11–19 days in the cat. The sporocysts in the dog were 13.5–15.4 x (10.6 ± 0.5) μm, and the sporozoites were 9.3 x 2.8 μm. He gave no more data. It would certainly be unusual if both the dog and cat were definitive hosts of the same species.

Toxoplasma gondii (Nicolle and Manceaux, 1908)
Nicolle and Manceaux, 1909

Synonyms. Leishmania gondii Nicolle and Manceaux, 1908; *Toxoplasma cuniculi* (Splendore, 1908); *Toxoplasma canis* Mello, 1910; *Toxoplasma talpae* von Prowazek, 1910; *Toxoplasma columbae* Yakimoff and Kohl-Yakimoff, 1912; *Toxoplasma pyrogenes* Castellani, 1913; *Toxoplasma musculi* Sangiorgi, 1913; *Toxoplasma sciuri* Coles, 1914; *Toxoplasma ratti* Sangiorgi, 1915; *Toxoplasma francae* (de Mello, 1915) Wenyon, 1926; *Toxoplasma caviae* Carini and Migliano, 1916; *Toxoplasma nikanorovi* Zasukhin and Gaisky, 1930; *Toxoplasma laidlawi* Coutelen, 1932; *Toxoplasma wenyoni* Coutelen, 1932; *Toxoplasma crocidurae* Galli-Valerio, 1933; *Toxoplasma fulicae* de Mello, 1935; *Toxoplasma hominis* Wolf, Cowen and Paige, 1939; *Toxoplasma gallinarum* Hepding, 1939, *Isospora gondii* (Nicolle and Manceaux, 1908) Tadros and Laarman, 1976.
Type Definitive Host. Domestic cat *Felis catus.*
Other Definitive Hosts. Jaguarundi *F. yagouaroundi,* ocelot *F. pardalis,* mountain lion *F. concolor,* Asian leopard cat *F. bengalensis,* bobcat *Lynx rufus,* probably cheetah *Acinonyx jubatus* (Mil-

ler, Frenkel, and Dubey, 1972; Jewell et al., 1972; Marchiondo, Duszynski, and Maupin, 1976).

Type Intermediate Host. Gondi *Ctenodactylus gundi.*

Other Intermediate Hosts. Over 200 species of mammals (including felids) and birds known.

Location. One type of meront, gamonts, gametes, zygotes, and oocysts in epithelial cells of villi of small intestine of the cat and other felids. Meronts and merozoites in many types of cells of intermediate host, including neurons, microglia, endothelium, liver parenchyma cells, lung and glandular epithelial cells, cardiac and skeletal muscle cells, fetal membranes, and leukocytes. In acute infections, merozoites may be found free in the blood and peritoneal exudate. Merozoites normally occur in the cytoplasm of host cells, but may on rare occasions invade the nucleus, at least in tissue culture cells (Remington, Earle, and Yagura, 1970).

Geographic Distribution. Worldwide.

Prevalence. Parenteral infections of intermediate hosts with *T. gondii* are common. To quote Levine (1973):

Toxoplasmiasis is apparently extremely common in man and also in many domestic animals. As Jacobs (1957) said, there is a sea of *Toxoplasma* infection around us. However, toxoplasmosis is far less common. Most infections are inapparent, and the disease itself appears only under special circumstances, many of which are still unknown.

Most of the surveys which have been made for *Toxoplasma* have been serologic and indicate either previous or present infections. Surveys in which the organism itself was isolated are more reliable, but much more time-consuming and expensive.

The prevalence of antibodies varies widely in man in different geographic locations. For instance, according to Jacobs (1957), there is relatively less infection in California than in the eastern United States. Walls, Kagan and Turner (1967) found a high prevalence among military recruits from the Appalachian area and a low one in the Rocky Mountain area. In Brazil, however, Walls and Kagan (1967) found that the prevalence of antibodies among military recruits was high in low plain areas and low in mountainous areas.

In seven surveys in the U.S., the dye test was positive in 4–35 percent (mean, 22 percent) of the people examined; it was positive in 10–68 percent (mean, 34 percent) in ten surveys in other countries. The above results give some idea of the range of positive reactions which may be expected in different surveys.

Lunde (1973) said that about 30% of the population in the United
States has antibodies to *T. gondii*.

In surveys for oocysts in cats, Shah (1970) found oocysts (which
might have been those of either *T. gondii*, *T. hammondi*, or *Besnoitia*) in the feces of 1.5% of 130 cats in Illinois, Guterbock and
Levine (1977) found them in 1% of 217 cats in Illinois, Dubey
(1973) found 0 in 515 cats (of which only 126 were over 6 months
old) in Kansas City, Christie, Dubey, and Pappas (1976) found
them in 1% of 1,000 cats in Ohio, Wallace (1971, 1973) in 1.4%
of 1,604 cats on Oahu, Hawaii, Fameree and Cotteleer (1976) in
2% of 252 cats in Belgium, and Nery-Guimaraes and Lage (1973)
in 1 of 185 cats in Brazil.

Using the indirect hemagglutination test on wild animals on a
sheep range at the Hopland Field Station, California, Franti et al.
(1975) found that 100% of 5 bobcats *Lynx rufus*, 40% of 10 coyotes *Canis latrans*, 44% of 9 gray foxes *Urocyon cinereoargentus*,
40% of 5 raccoons *Procyon lotor*, 33% of 6 striped skunks *Mephitis
mephitis*, and 17% of 12 cats *Felis catus* were positive.

Oocyst Structure. Oocysts are produced in the intestinal epithelial cells of felids. The oocysts in the feces are not sporulated when
passed. They are spherical at first, but after sporulation are subspherical, 11–14 x 9–11 (mean, 12.5 x 11) μm, without a micropyle, residuum, or polar granule, and contain 2 ellipsoidal sporocysts about 8.5 x 6 μm, without a Stieda body but with a residuum.
Each sporocyst contains 4 sporozoites about 8 x 2 μm.

Sporulation. 2–3 days at 24 C, 5–8 days at 15 C, and 14–21 days
at 11 C (Frenkel, Dubey, and Miller, 1970). Oocysts do not sporulate at 4 or 37 C (Dubey, Miller, and Frenkel, 1970a).

Merogony. Animals become infected by ingesting sporulated
oocysts or infected meat or animals, or congenitally via the pla-
centa. Congenital toxoplasmosis of the newborn resulting from infection of the mother while she is pregnant is probably the most
common form in man and perhaps sheep, but it is not nearly so
important in cats (Dubey and Hoover, 1977; Dubey, 1977). Mice
can be infected congenitally for generation after generation; Beverly
(1973), for instance, reported that toxoplasmosis had been transmitted congenitally through at least 9 successive generations in
Swiss mice. Experimental infections can be established by intra-
venous, intraperitoneal, or any other type of parenteral inoculation

or by feeding. Following experimental inoculation the protozoa proliferate for a time at the site of injection and then invade the blood stream and cause a generalized infection. Susceptible tissues all over the body are invaded, and the parasites multiply in them, causing local necrosis. The parasitemia continues for some time, until antibodies appear in the plasma, after which the parasites disappear from the blood and more slowly from the tissues. They finally remain only in meronts ("cysts"), and only in the most receptive tissues. In general, the spleen, lungs, and liver are cleared of parasites relatively rapidly, the heart somewhat more slowly, and the brain much more slowly. Residual infections may persist for a number of years. Dubey (1977a), for example, found that this organism remained in the tissues of cats for at least 143–473 days after the cats had ingested meronts containing bradyzoites. He found meronts in the heart or intestinal mucosa of 5 out of 7 cats, in the skeletal muscles, diaphragm, and spinal cord of 4, in the brain, spleen, and kidneys in 3, and in other organs of 1 or 2.

The sex of merozoites is not predetermined. Pfefferkorn, Pfefferkorn, and Colby (1977) found that the offspring of a single merozoite (produced by cloning in tissue culture) were able to produce both micro- and macrogametes and viable oocysts.

When the sporulated oocyst is ingested by a susceptible animal (which may be almost any mammal or one of a considerable number of birds), the sporozoites emerge and pass to the parenteral tissues via the blood and lymph; any type of cell may be invaded. Here they multiply by endodyogeny. The stage in which this occurs has been called a pseudocyst, terminal colony, colony aggregate stage, or group stage; the last term is preferable. The merozoites within it are tachyzoites ("fast", i.e., rapidly developing zoites; a term introduced by Frenkel, 1973); they have also been called proliferative forms and endozoites (a term first used by Sénaud, 1963). They multiply by endodyogeny. The group stage with its tachyzoites is the stage found in the leukocytes in peritoneal exudate, but it also occurs in other parenteral locations, such as the liver, lungs, and submucosa; this is the stage occurring in acute toxoplasmosis. (We prefer the term tachyzoite because the "endozoites" are not necessarily *in* anything.)

There is an indefinite number of generations of tachyzoite. Eventually they enter other cells and induce the host cell to form a wall around them, forming the structure generally called a cyst; actually, it is a pseudocyst or meront. Within it a large number of bradyzoites ("slow" zoites, i.e., slowly developing zoites, a term introduced by Frenkel, 1973) is formed by endodyogeny. Bradyzoites are also called cystozoites, a term used by Vivier (1970), or cyst forms; however, this term is misleading because they do not occur in a true cyst. The meronts and bradyzoites are much more resistant to trypsin and pepsin than the tachyzoites, and they may remain viable in the tissues for years. This is the stage found commonly in the brain, but it also occurs in other tissues such as muscle; it is the stage found in chronic infections.

The above meront is the end of the life cycle in all animals except felids so far as is known. In the cat and other felids the bradyzoites enter the intestinal tissues and multiply. Dubey and Frenkel (1972a) recognized 5 types of multiplicative stages in the intestinal epithelial cells: Type A is present 12–18 hours after feeding meronts and multiplies by endodyogeny. Type B is present 12–54 hours after feeding meronts and multiplies by endogenesis (endodyogeny and endopolygeny) (Piekarski, Pelster, and Witte, 1971; Vivier, 1970). Type C is present 28–54 hours after feeding meronts and multiplies by ordinary schizogony. Type D is present 32 hours to 15 days after feeding meronts and multiplies by schizogony, endopolygeny, or splitting. Type E is present 3–15 days after feeding meronts and multiplies by schizogony. It is not known to how many generations these types belong.

The merozoites are 5–8 x 1–2 μm and have a 3-membraned complex at the surface, each membrane consisting of 2 electron-dense layers separated by electron-light material. They have an apical complex consisting of: (1) 2 polar rings at the anterior end (and a similar ring at the posterior end); (2) a short, truncate hollow conoid 0.2–0.35 x 0.15–0.35 μm composed of 6–7 microtubules spirally coiled at an angle of 45°–50°; (3) 20 to perhaps 30 (according to some, 5–9) cylindrical or club-shaped rhoptries of variable length, 0.02 μm in diameter after leaving the conoid and thickening toward the posterior to form club-shaped or sausage-shaped structures 0.08–0.2 μm in diameter, some extending nearly to the posterior end and others not reaching the level of the nucleus; they become very slender and tortuous as they approach the conoid and are apparently open to the outside at the anterior end after passing through the conoid (Vivier and Petitprez, 1972, said they contain mucopolysaccharides and sometimes acid phosphatase); (4) about 50 curved, rodlike micronemes (which may be associated with or be stages in the development of the rhoptries) anterior to the nucleus; (5) 22 longitudinal subpellicular microtubules arising from a ring at the level of the conoid and running toward the posterior about 1/5–2/3 of the body length. Just in front of the nucleus is the Golgi apparatus. There are 1 or more micropores in the pellicle. The cytoplasm is somewhat vacuolated and contains numerous ribosomes, rough endoplasmic reticulum, and 1 to several mitochondria. The nucleus is usually spherical or ovoid, about 1–2 μm in diameter, and contains a large nucleolus.

Gametogony. The male and female gamonts are in the intestinal epithelial cells. The microgamonts produce 12–32 slender, crescentic microgametes about 3 μm long, which have 2 flagella plus the rudiments of a third (Pelster and Piekarski, 1971). The macrogamonts are presumably haploid and simply grow; they are really macrogametes already. Fertilization takes place and the resultant zygotes form walls around themselves, become oocysts, and are released into the intestinal lumen.

Prepatent Period. In cats this period is 3–5 days after feeding parenteral meronts ("cysts"), 7–10 days after feeding merozoites, and 20–24 days after feeding fecal oocysts (Frenkel, Dubey, and Miller, 1970; Dubey, Miller, and Frenkel, 1970b), 2–7 days after feeding "cysts" from mice, and 7 days or more after feeding oocysts from cats (Witte and Piekarski, 1970).

Patent Period. 1–2 weeks in cats.

Pathogenicity. The sexual stages of *T. gondii* are apparently not pathogenic for felids. The parenteral stages may or may not cause symptoms. Toxoplasmosis may vary from an inapparent infection to an acutely fatal one. Asymptomatic toxoplasmiasis is the most common.

In humans, a common form of the disease is the congenital type found in newborn infants. It is characterized by encephalitis, rash, jaundice, and hepatomegaly, usually associated with chorioretinitis, hydrocephalus, and microcephaly; the mortality rate is high (Feldman, 1953; Feldman and Miller, 1956). Congenital toxoplasmosis occurs in 0.25–7.0/1,000 live births in different countries. About 3,000 babies are born each year, presumably in the United States, with toxoplasmosis; 5%–15% of them die, 8%–10% have marked brain and eye lesions, 10%–13% have moderate to marked visual damage, and 58%–72% are clinically normal at birth but may develop retinochoroiditis in childhood or young adulthood (Frenkel, 1973). The total cost of neonatal human toxoplasmosis in the United States was estimated by Frenkel (1973) to be about $31–$40 million, including costs for hospitalization, institutionalization, and special education.

Acquired (i.e., noncongenital) human toxoplasmosis has many different manifestations. Siim (1956) divided them into 4 main types. The most common is characterized by lymphadenopathy; it may be febrile, nonfebrile, or subclinical. The second type is a typhuslike exanthematous disease. The third type is a cerebrospinal form, generally fatal, but fortunately rare. The fourth type is an ophthalmic form characterized by chronic chorioretinitis.

Remington, Jacobs, and Kaufman (1960) reviewed toxoplasmosis in the human adult.

Toxoplasmosis is similar in domestic animals to the disease in humans. It has been reported in the dog, cat, pig, ox, sheep, squirrel, monkey, marmoset, and chicken, among other animals. Perinatal mortality is especially important in sheep (Hartley and Marshall, 1957; Hartley and Kater, 1963, 1964). Levine (1973) and Beverley (1974) gave further details.

Immunity. Animals that have had toxoplasmosis or toxoplasmiasis are immune to reinfection. Various serologic tests have been recommended for the diagnosis of infections due to *Toxoplasma.* These include the Sabin-Feldman dye test (Sabin and Feldman, 1948; Sabin et al., 1952), complement fixation test (Warren and Sabin, 1943; Sabin, 1949), hemagglutination test (Jacobs and Lunde, 1957), skin test (Frenkel, 1948, 1949), fluorescine-labeled antibody test (Goldman, Carver, and Sulzer, 1957), and latex agglutination test (Lunde and Jacobs, 1967). The dye test and hemagglutination test are perhaps the most widely used.

T. gondii does not cross-react serologically with other coccidia (see Piekarski and Witte, 1971; Aryeetey and Piekarski, 1976).

Cross-Transmission Studies. Many animals are readily susceptible to parenteral infection with *T. gondii*—a fact advantageous in diagnosis. The most certain method of diagnosis is by isolation of the parasites themselves, and this is done by inoculation of experimental animals, usually mice. Eichenwald (1956) considered mice, golden hamsters, and guinea pigs the most sensitive animals in his experience. Simitch, Petrovitch, and Bordjochki (1956) preferred the ground squirrel *Spermophilus citellus*, while Lainson (1957) recommended the multimammate rat *Rattus coucha*.

The only animals known to pass oocysts are Felidae. Janitschke and Werner (1972), for instance, found that oocysts would develop in *F. bengalensis* but not in the viverrids *Vivera zibethica, Paguma larvata,* or *Herpestes auropunctatus*. Jewell et al. (1972) found that oocysts would develop in *F. yagouarondi* and *F. pardalis* but not in the procyonids *Nasua nasua, Potos flavus,* or *Bassaricyon gabbii*. Miller, Frenkel, and Dubey (1972) found that oocysts were shed by the bobcat *L. rufus*, mountain lion *F. concolor*, and Asian leopard cat *F. bengalensis* after they had been fed *T. gondii* meronts, but not by 16 nonfeline mammals of 7 orders and 9 birds of 5 orders.

Cultivation. *T. gondii* is readily cultivated in tissue culture and chicken embryos. It was first cultivated in both by Levaditi et al. (1929) and has since been cultivated by many persons (see Doran, 1973).

Remarks. This is a much abbreviated review of *T. gondii* infections. The literature on this coccidium is vast. Jira and Kozojed (1970) published a two-volume bibliography with 7,763 entries for the period 1908–67, and many papers have been published since that time. Various aspects of the disease have been reviewed since 1970 by Jacobs (1970), Frenkel (1970, 1973, 1974, 1974a), Kozar (1970), Hutchison et al. (1971), Piekarski and Witte (1971), Jones (1973), Levine (1973, 1977b), Siim (1974), Senaud and Mehlhorn (1975), and Beverley (1976), among others. That *T. gondii* is a coccidium having an oocyst in cats like that of *Isospora* was discovered independently by Overdulve (1970) in the Netherlands, Work and Hutchison (1969, 1969a), and Hutchison et al. (1970) in Scotland and Denmark, Weiland and Kühn (1970) in Germany, Sheffield and Melton (1970) in Maryland, Frenkel, Dubey, and Miller (1970) in Kansas, and G. D. Wallace (unpubl.) in Hawaii. It has been amply confirmed since then.

Toxoplasma hammondi (Frenkel, 1974) Levine, 1977

Synonyms. Hammondia hammondi Frenkel, 1974, Frenkel and Dubey, 1975; *Isospora hammondi* (Frenkel and Dubey, 1975), Tadros and Laarman, 1976.
Type Definitive Host. Domestic cat *Felis catus.*
Other Definitive Hosts. None known.
Intermediate Host. House mouse *Mus musculus* (experimental).
Other Intermediate Hosts. The following experimental infections were successful: laboratory rat *Rattus norvegicus,* golden hamster *Mesocricetus auratus,* guinea pig *Cavia porcellus* (Frenkel and Dubey, 1975, 1975a), domestic dog *Canis familiaris* (Dubey, 1975b), marmoset *Saguinus nigricollis* (Dubey and Wong, 1978). In addition, Frenkel and Dubey (1975, 1975a) infected the white-footed mouse *Peromyscus leucopus* and the multimammate rat *Mastomys coucha* as judged by the development of antibodies, but apparently did not obtain oocysts by feeding these animals to cats. Dubey and Wong (1978) obtained a dye test titer of 1:16 in the monkey *Macaca fascicularis, M. mulatta,* and *Cercocebus* by feeding oocysts from cats, but the monkey tissues were not infective for cats.
Location. One type of meront, gamonts, gametes, and oocysts in epithelial cells of villi of small intestine of the cat. Meronts and merozoites mostly in striated muscles of mice, but a few in the brain and to a much lesser extent in the spleen, liver, lungs, and mesenteric lymph nodes (Frenkel and Dubey, 1975; Dubey and Streitel, 1976b). Rommel and von Seyerl (1976) found meronts only in the striated muscles.
Geographic Distribution. Presumably worldwide. So far, however, reported only from North America (Kansas, Frenkel and Dubey, 1975, 1975a) and Europe (Germany, Rommel and von Seyerl, 1976).
Prevalence. Unknown.
Oocyst Structure. The oocysts are not sporulated when shed in cat feces. They are subspherical to spherical, 11–13 x 10–13 (mean, 11 x 11) μm, with a 2-layered, smooth, colorless wall about 0.5 μm thick. The sporulated oocysts are subspherical to ellipsoidal, 13–14 x 10–11 (mean, 13 x 11) μm, without a micropyle, residuum, or polar granule. The sporocysts are ellipsoidal, 9–11 x 6–8 (mean, 10 x 6.5) μm, without a Stieda body but with a residuum. The sporo-

zoites are elongate and curved, about 7 x 2 μm, with a nucleus near the center (Frenkel and Dubey, 1975). (Rommel and von Seyerl, 1976, said that the unsporulated oocysts were 12–18 x 10–15 μm with a mean of 14 x 12 μm.)

Sporulation. 3 days at 20–23 C in 2% sulfuric acid solution (Frenkel and Dubey, 1975) or 2–3 days at 21 C (Rommel and von Seyerl, 1976).

Merogony. Both group stages containing tachyzoites and meronts containing bradyzoites have been found; tachyzoites in mice are not infective for cats, but bradyzoites are (Dubey and Streitel, 1976b). Meronts in the skeletal muscles of mice were found by Frenkel and Dubey (1975) to be 100–340 x 40–95 μm. They were nonseptate, had no radial spines, and contained slender merozoites only. Rommel and von Seyerl (1976) found meronts 44–100 x 32–45 μm in the skeletal muscles of mice 6–8 weeks after they had been fed sporulated oocysts.

Merozoites from tissue culture were studied with the electron microscope by Sheffield, Melton, and Neva (1976). They have a typical coccidian fine structure, with a double-membraned pellicle, subpellicular microtubules, conoid, micronemes, rhoptries, and a micropore, all resembling those of *T. gondii*. The only difference was that the *T. hammondi* merozoites contained storage granules that appeared to be depleted after cell invasion.

Endodyogeny was the only type of division seen by Sheffield, Melton, and Neva (1976).

As in *T. gondii*, there is a third wave of multiplication in the epithelium of the small intestine before gametogony (Frenkel and Dubey, 1975).

Gametogony. This process takes place in the wall of the cat small intestine, but we have no details regarding it.

Prepatent Period. The prepatent period after feeding infected mice to cats is 5–8 days (Frenkel and Dubey, 1975; Dubey, 1975b), 6 days (Dubey and Streitel, 1976b), or 6–16 (mean, 9) days (Rommel and von Seyerl, 1976). Meronts containing bradyzoites become infective for cats 10 days after mice have been fed oocysts from the cat (Dubey and Streitel, 1976b). The prepatent period in cats after they have been fed the tissues of dogs that had been fed sporulated oocysts 16 and 38 days before was found by Dubey (1975b) to be 7–9 days.

Patent Period. This was found by Frenkel and Dubey (1975) to be 10–28 days or by Dubey (1975b) to be 1–2 weeks in cats. The cats reshed oocysts later.

Pathogenicity. T. hammondi is apparently not pathogenic for cats. According to Rommel and von Seyerl (1976), mice fed sporulated oocysts developed a severe anorexia with apathia after 7–8 days, and some of them died.

Immunity. Cats that have ceased shedding oocysts may reshed them later. Dubey (1975b) found that the prepatent period was not affected by hyperadrenocorticism caused by weekly injections of 6-methyl prednisolone acetate, but this drug did cause cats to reshed oocysts. He concluded that immunity to *T. hammondi* in cats was less stable than immunity to *T. gondii.*

Some intermediate hosts were found by Frenkel and Dubey (1975, 1975a) to have low levels of antibody and some cross-immunity against *T. gondii,* but they found neither in infected cats. Rommel and von Seyerl (1976) found that of 12 histologically positive mice, 1 was negative to the Sabin-Feldman dye test for *T. gondii,* 1 had a titer of 1:16, 4 of 1:64, and 6 of 1:256. Under similar circumstances, *T. gondii* would have elicited titers of 1:16,000 to 1:64,000.

Cross-Transmission Studies. T. hammondi cannot be transmitted from cat to cat by means of oocysts, or from mouse to mouse by injection or feeding of tissues or suspensions of bradyzoites (Frenkel and Dubey, 1975, 1975a; Rommel and von Seyerl, 1976). In other words, *T. hammondi* has an obligate 2-host life cycle. There is no evidence that *T. hammondi* is transmitted transplacentally from mother cats to their kittens (Dubey, 1977).

Oocysts are not produced in the dog, although an intermediate stage infective for cats occurs in its parenteral tissues (Dubey, 1975b). Dubey and Streitel (1976b) found that *T. hammondi* oocysts are not infectious for chickens. Fayer and Frenkel (1979) could not infect calves either as intermediate, transport, or definitive hosts.

Cultivation. Sheffield, Melton, and Neva (1976) cultivated *T. hammondi* in monkey kidney, mouse embryo, and the WI-38 strain of human diploid fibroblast tissue cultures. They observed division daily for 8 days. Endodyogeny was the only type of division they saw.

Toxoplasma pardalis (Hendricks et al., 1979) n. comb.

Synonyms. Hammondia pardalis Hendricks et al., 1979; *Isospora* sp. Long and Speer, 1977.

Type Definitive Host. Ocelot *Felis pardalis* L.

Other Definitive Hosts. The following are experimental definitive hosts: Domestic cat *F. catus,* jaguarundi *F. yagouaroundi.*

Intermediate Host. House mouse *Mus musculus* (experimental).

Location. Intestinal contents of definitive hosts; mesenteric lymph nodes, lungs, and intestinal submucosa of mouse.

Geographic Distribution. Panama.

Prevalence. This species was found in the feces of a single ocelot at the Air Force Survival School Zoo, Albrook Air Force Station, Panama Canal Zone.

Oocyst Structure. Ovoid, 36–46 x 25–35 (mean, 41 x 28.5) μm, with 1-layered, colorless, smooth to slightly rough wall 1.8 μm thick, lined by dark membrane, with micropyle (?) or suture line at small end and similar structure that does not dissolve on excystation at the large end, usually without residuum, with polar granule. Sporocysts ellipsoidal, 19–25 x 14–19 (mean, 22 x 16) μm, without Stieda body or substiedal body, with 1 to several residual globules, with wall about 1 μm thick. Sporozoites sauage-shaped, usually lying lengthwise in sporocysts, without clear globule.

Sporulation. 5–7 days at room temperature in 2.5% potassium bichromate solution.

Merogony. In mice fed 5,000 oocysts 10 days previously, Hendricks et al. (1979) found meronts and tachyzoites in the lymphoid tissue of the colon and alveolar septa of the lungs; the merozoites were 6 x 3 μm. At 12 days they found small meronts 8–10 x 6–8 μm with immature merozoites in the alveolar septa of the lungs; some of the organisms were within histiocytes. At 15 days they found meronts and tachyzoites in the lymphoid tissue of the colon and alveolar septa of the lungs; the meronts in the lymph nodes were 13–16 x 10–13 μm. They found no tissue cysts in mice intraperitoneally infected 30 days before with material from a donor infected mouse.

Gametogony. Unknown.

Prepatent Period. 5–8 days in felids.

Patent Period. 5–13 days in felids.

Pathogenicity. This species is apparently not pathogenic for the felid definitive hosts. However, it is for house mice. Hendricks et al. (1979) found that most of the mice fed 100,000 or 500,000 oocysts became sick within 1 week and 63% of 40 were dead within 15 days.

Cross-Transmission Studies. As indicated above, Hendricks et al. (1979) were unable to infect mice with stages from other mice. They were also unable to infect raccoons. They infected mice from the initial ocelot and then 5 ocelots, 6 domestic kittens, and a jaguarundi by feeding them infected mice.

Remarks. This is presumably the *Isospora* sp. illustrated with Zeiss-Nomarski interference-contrast photomicrographs by Long and Speer (1977) in their review paper on excystation and invasion of host cells by coccidia. They received the oocysts from Dr. Hendricks.

Toxoplasma spp. Entzeroth, Scholtyseck, and Greuel, 1978

Entzeroth, Scholtyseck, and Greuel (1978) fed esophagus, diaphragm, and peritoneal muscles of roe deer *Capreolus capreolus* to a cat and found unsporulated oocysts 12 x 11 μm in its feces 8 days later. After sporulating these oocysts and feeding them to mice, they found typical *"Toxoplasma"* cysts in their brains and typical *"Hammondia hammondi"* cysts in their muscles. After a prepatent period of 11 days, they also found unsporulated oocysts 13.1 x 11.6 μm resembling those of "the small form of *Isospora bigemina (Hammondia)* described by Heydorn (1973) and Rommel and Seyerl (1976) in the feces of the fox." These oocysts did not infect mice.

Besnoitia besnoiti (Marotel, 1913) Henry, 1913

Synonyms. Sarcocystis besnoiti Marotel, 1913; *Gastrocystis besnoiti* (Marotel, 1912) Brumpt, 1913; *Gastrocystis robini* Brumpt, 1913; *Globidium besnoiti* (Marotel, 1912) Weynon, 1926; *Isospora besnoiti* (Marotel, 1912) Tadros and Laarman, 1976.

Definitive Host. Domestic cat *Felis catus,* wild cat *F. libyca* (see Peteshev, Galuzo, and Polomoshnov, 1974).

Intermediate Hosts. Ox *Bos taurus,* zebu *B. indicus.*

Other Intermediate Hosts. Goat *Capra hircus* (Bwangamoi, 1968); blue wildebeest *Connochaetes taurinus,* impala *Aepyceros melampus,* and kudu *Tragelaphus strepsiceros* (McCully et al., 1966). Natural infections have been found in the above. In addition, experimental infections have been produced in the following species by injection of meronts or bradyzoites from infected cattle or antelope: domestic rabbit *Oryctolagus cuniculus* (Pols, 1954, 1960), domestic sheep *Ovis aries* (Pols, 1954, 1960), gerbil *Meriones tristrami shawi* (Neuman, 1962a), golden hamster *Mesocricetus auratus* (Neuman and Noble, 1963; Bigalke, 1968), ground squirrel *Spermophilus fulvus* (Peteshev, Polomoshnov, and Eshtokina, 1975), marmot *Marmota* sp. (Peteshev, Galuzo, and Polomoshnov, 1974), and house mouse *Mus musculus* (Peteshev, Galuzo, and Polomoshnov, 1974; Rommel, 1975).

Location. Gamonts, gametes, zygotes, oocysts, and probably the last generation of meronts are presumably in the intestine of felids. The meronts are in the cutis, subcutis, scleral conjunctiva, connective tissue, fascia, serosae, mucosae of the nose, larynx, trachea, and other places of intermediate hosts. Merozoites are in the blood, either extracellularly or in monocytes, and in smears of lymph nodes, lungs, testes, etc. of intermediate hosts.

Geographic Distribution. Europe (southern France, Pyrenees, Portugal), USSR (Kazakhstan), Asia (China, Israel), Africa (Angola, Botswana, Congo, Burundi, Kenya, Ruanda, South Africa, Sudan, Tanzania, Uganda, Zambia), South America (Venezuela) (see Neuman, 1962, 1972; Er-Hsiang, 1957; Vsevolodov, 1961; Bwangamoi, 1968; Pols, 1954, 1960; Hussein and Haroun, 1975).

Prevalence. According to Hofmeyr (1945), *B. besnoiti* is endemic in South Africa throughout the whole of the bushveld area from the Western Transvaal to Potgietersrust district and probably further north. It is highly enzootic in the Transvaal and extends south into the Orange Free State (Bigalke and Naudé, 1962; Bigalke and Schoeman, 1967; Bigalke, 1968). Hérin (1952) found it in about 2% of the cattle he examined in Ruanda-Urundi. Khvan (1969) found it in as many as 50% of some herds of cattle in southern and southeastern Kazakhstan. Neuman (1972) found that 56% of 1,411 beef and dairy cattle in Israel had positive immunofluorescence test antibody titers.

Oocyst Structure. The oocysts are similar to those of *Isospora.* They are ovoid and unsporulated when passed, 14–16 x 12–14 μm, apparently without a micropyle. After sporulation they contain 2 sporocysts, each with 4 sporozoites, and resemble the oocysts of *Toxoplasma* (Peteshev, Galuzo, and Polomoshnov, 1974). No further structural information appears to be available.

Sporulation. Apparently unknown; 7–8 days or less.

Merogony. In cattle the first-generation meronts are found in the endothelial cells of the blood vessels. The merozoites produced here enter other cells and form large meronts (pseudocysts) with a thick wall containing several host cell nuclei and thousands of merozoites but no metrocytes.

In experimentally infected rabbits, Pols (1954) saw the "initial" stages as early as 16–18 days after inoculation. When a zoite invades a histiocyte, a vacuole is formed around it. The zoites in tissue sections are about 3 x 1.5 μm, and the vacuoles are about 8 μm in diameter. The zoites multiply by binary fission. Pols saw a few cases of multiple fission, but they were so rare that he considered them to be aberrant. The zoites may actually divide by endodyogeny, since this occurs in *B. jellisoni* (Scholtyseck, Mehlhorn, and Müller, 1973).

The host cell nucleus begins to divide at the same time that the zoites do, forming a multinucleate cell. As the parasites multiply within the vacuole, the vacuole becomes larger, and the host cell cytoplasm is compressed to form a narrow rim. This is the middle layer of the pseudocyst (meront) wall. Within it is an inner membrane, which can be seen only if 2 zoites have invaded the same host cell, in which case it forms a thin line between the resultant meronts; it is uncertain whether it is formed by the parasite, the host, or both. Concentric layers of collagenous fibers are laid down around the host cell to form a hyaline capsule around the whole; this is the outer layer of the pseudocyst.

The mature pseudocysts (meronts) in cattle are more or less spherical, without septa, and about 100–500 μm in diameter. Their wall is composed of a thin inner layer containing several flattened giant nuclei and a thick, homogeneous or concentrically laminated outer wall; both these layers are formed by the host. The merozoites in the pseudocysts are crescentic or banana-shaped, with one end pointed and the other rounded.

According to Pols (1954), the merozoites in smears of the blood, lung, and testis of experimentally infected rabbits are 5–9 x 2.5 μm and are usually elongate oval and slightly pointed at one end. Banana-shaped and crescentic forms are found more rarely. The nucleus is more or less central.

The fine structure of the merozoites has apparently not been investigated. However, Scholtyseck, Mehlhorn, and Müller (1973) found that there are no metrocytes in the closely related *B. jellisoni,* and that its merozoites are 7–9 x 2–3 μm and have the typical apicomplexan structure, with 22 subpellicular microtubules, 80–100 micronemes, 3–5 rhoptries, a conoid, a micropore, and 20–30 enigmatic bodies (membrane-bound, electron-light, spindle-shaped bodies with an electron-dense core whose function is unknown).

Gametogony. This process has apparently not been studied. Presumably it occurs in the intestinal cells of the cat.

Prepatent Period. Oocysts appear in the feces of cats 16 days (Peteshev, Galuzo, and Polomoshnov, 1974) or 4–25 days (Rommel, 1975) after they have been fed meronts (pseudocysts) from cattle. They were found in the feces of kittens by Peteshev, Galuzo, and Polomoshnov (1974) 6–7 days after they had been fed the brains of 2 infected mice.

Patent Period. 3–15 days in cats (Peteshev, Galuzo, and Polomoshnov, 1974).

Pathogenicity. B. besnoiti is apparently not pathogenic for the cat definitive host. However, it may cause serious disease in the intermediate host. Most infected cattle have low grade, chronic infections without skin lesions. The typical signs in clinically affected cattle are anasarca followed by scleroderma, alopecia, and seborrhea (Bigalke and Naudé, 1962). The most complete description of clinical bovine besnoitiosis was given by Hofmeyr (1945). He found it in cattle of all ages from 6 months up. Aged animals were also affected, He recognized 3 stages in the course of the disease. As described by Levine (1973) these were:

The febrile stage. The first sign of besnoitiosis is fever, up to 107 F but usually lower. The animal develops photophobia and stays in the shade. The hair loses its luster, especially along the buttocks, limbs, flanks, lower abdomen, and neck. Marked anasarca develops, especially along the lower line but sometimes over the whole body. The swellings are warm and tender. The animals have a stiff gait and are reluctant to move. The pulse is fast, respiration is rapid, and rumination decreases or ceases. Diarrhea is sometimes present, and abortions are not uncommon. The lymph nodes, especially the prescapular and precrural ones, are enlarged. Lachrymation and hyperemia of the sclera are present. The cornea is studded with whitish, elevated specks which are *Besnoitia* cysts. The nasal mucosa becomes bright red and is also studded with cysts. The mucosa may be swollen and there may be a rapidly progressive rhinitis; it starts with a mucous discharge which later becomes thick, hemorrhagic, and mucopurulent, forming dark brown crusts in the nostrils. If the pharynx and larynx are involved, there is a short cough. This stage may last five to ten days. The acute stage then subsides and the second stage begins.

The depilatory stage. In this stage, the pathologic and clinical pictures are dominated by skin lesions. The skin becomes greatly thickened and loses its elasticity. The hair falls out over the swollen parts, and the skin on the flexor surfaces cracks and a sero-sanguinous fluid oozes out. Necrosis of the skin develops on the parts in contact with the ground when the animal lies down. Toward the end of this stage, hard sitfasts develop on the sides of the stifles, brisket, and elbows. The anasarca subsides, leaving the skin with typical, broad wrinkles along the lower line. The photophobia decreases, and grazing is resumed in many cases. Death may occur at this stage. If not, the stage lasts two weeks to about a month and gradually passes into the third stage.

The seborrhea sicca stage. In this stage, most of the hair on the previously anasarcous skin has been lost, and the denuded parts are covered by a thick, scurfy layer. The sitfasts crack away from the underlying tissues, fissures remain in the flexor surfaces, the skin hardens, and deep scars show plainly. The hide resembles that of an elephant, and the animal looks as though it has mange. The lymph nodes are permanently enlarged, the protozoan cysts remain, and the animal is listless and debilitated.

In light infections in which there has been little hair loss, the animals become practically normal in appearance, but in more severe cases recovery requires months or even years, and the changes in the cutis and subcutis and the loss of most of the hair are permanent. In convalescent animals the remaining hair forms patterns resembling the markings on a giraffe.

The morbidity in a herd is 1 to 20 percent, and the mortality is about 10 percent.

Basson, McCully, and Bigalke (1970) reported that bovine strains of *B. besnoiti* in the acute stage produced degenerative and necrotic lesions, vasculitis, and thrombosis, mainly of the medium and smaller veins and some arteries. These coincided with parasitization of endothelial and other cells of the vessels, where the organisms multiplied before the beginning of the pseudocyst stage. These basic lesions are responsible for edema, degenerative changes, and even infarction, particularly in the testes and skin. Other characteristic features are a histiocytic reaction and mild eosinophil infiltration.

The pseudocyst stage in cattle apparently develops in enlarged histiocytes. Basson, McCully, and Bigalke (1970) could recognize them 11 days after inoculation. The host cells become multinuclear and form the pseudocyst wall. The pseudocysts become mature 71 days after inoculation.

Antelopes are not severely affected by *B. besnoiti.*

Basson, McCully, and Bigalke (1970) found that the bovine strains were markedly pathogenic, causing severe testicular or skin lesions; pseudocysts were rare. They found that the wildebeest and impala strains were only very mildly pathogenic for rabbits, but passage during the acute stage increased their pathogenicity. Lesions were usually confined to the internal tissues and organs, such as the myocardium, gut, and lungs, and pseudocysts were rare.

Peteshev, Galuzo, and Polomoshnov (1974) said that marmots died of besnoitiosis 10–12 days after having been inoculated with material from cattle. Merozoites were found in all their organs. They said that sheep and goats had a temperature of 41–42 C beginning 10–13 days after inoculation and lasting 3–9 days. During this time single merozoites were found in the blood. The temperature then declined, and the animals remained clinically healthy. They saw pseudocysts in the sclera of the eyes of 1 goat and 1 sheep for 2 months after inoculation. They saw no clinical signs in white mice.

Peteshev, Polomoshnov, and Eshtokina (1975) said that the ground squirrel *S. fulvus* was easily infected by parenteral inoculation with blood of heavily infected cattle, of merozoites obtained from pseudocysts, or orally by pseudocyst-containing tissues. The incubation period was 3 days, and all organs were infected; the animals died after 8–14 days. They recommended this ground squirrel for diagnosis.

Immunity. Animals infected with *B. besnoiti* are positive to the complement fixation test (Peteshev, Galuzo, and Polomoshnov, 1974), the indirect fluorescence antibody and ELISA (peroxidase) tests (Weiland and Kaggwa, 1976). There is no cross-reaction with *Toxoplasma gondii.*

Bigalke et al. (1973) and Bigalke, Schoeman, and McCully (1974) immunized cattle and rabbits against *B. besnoiti* by use of a live vaccine prepared from a strain from the blue wildebeest grown in tissue culture. Remington (1969) found that infection of mice with *B. besnoiti* protected them against *Listeria monocytogenes, Salmonella typhimurium, Brucella melitensis, Plasmodium berghei, T. gondii,* and Mengo virus. He considered that common mechanisms of intracellular immunity exist and that the macrophage system is the effector arm of the observed resistance.

Cross-Transmission Studies. Only *F. catus* and *F. libyca* have so far been found to produce oocysts. Rommel (1975) found that oocysts were not produced in the dwarf vulture *Pseudogyps africanus* and maribou stork *Leptoptilus crumeniferus.* Susceptible intermediate hosts have been listed above. Peteshev, Galuzo, and Polomoshnov (1974) were unable to produce infections in the wolf, corsac fox, hedgehog, or rook.

Cultivation. B. besnoiti is readily cultivated in tissue culture. Bigalke (1962), for example, used lamb and calf kidney monolayer cell cultures. Neuman (1974) cultivated it in sheep thyroid monolayer cell cultures, transferring it every 5–10 days for 4.5 years. He found that it grew very slowly and did not survive for more than 6 passages in baby hamster kidney or HeLa cells. Akinchina and Doby (1969) used rabbit and pig kidney cell cultures. The vaccine prepared by Bigalke, Schoeman, and McCully (1974) was grown in primary lamb kidney or an established line of green monkey kidney cell (Vero) tissue culture.

Besnoitia wallacei (Tadros and Laarman, 1976) Dubey, 1977

Synonym. Isospora wallacei Tadros and Laarman, 1976.
Type Definitive Host. Domestic cat *Felis catus.*
Intermediate Hosts. The following are all experimental intermediate hosts: laboratory (Norway) rat *Rattus norvegicus,* house mouse *Mus musculus,* Polynesian rat *R. exulans.*
Location. Meronts and merozoites occur in the small intestine (lamina propria, adjacent endothelial cells, or intimal cells) and liver

of cats. Gamonts, gametes, zygotes, and oocysts are formed in the goblet cells of the small intestine epithelium of the cat. Meronts ("cysts") and bradyzoites occur in the heart, mesentery, small intestine wall, and to a lesser extent in the liver and lungs of rats and mice.

Geographic Distribution. This species has so far been found only in Hawaii.

Prevalence. Unknown.

Oocyst Structure. Unsporulated oocysts subspherical, 16–19 x 10–13 (mean, 17 x 12) μm, with a 2-layered, smooth, light brown to pink wall 0.5 μm thick. Sporulated oocysts ellipsoidal to subspherical, 15–18 x 12–15 (mean, 16 x 13) μm, without micropyle, residuum, or polar granule. Sporocysts ellipsoidal, 11 x 7–8 (mean, 11 x 8) μm, without Stieda body, with diffusely distributed residual granules. Sporozoites about 10 x 2 μm within sporocyst (and 8 x 2.5 μm outside it when stained with Giemsa), lying lengthwise in sporocysts.

Sporulation. 64–96 hours at 24 C in 1%–2% sulfuric acid solution when agitated and exposed to air (Frenkel, 1977).

Merogony. Frenkel (1975) described merogony of this species. Following the feeding of sporulated oocysts to rodents, meronts containing tachyzoites were not found by Frenkel (1977), but it is possible that they were present, since a single transfer of 6-day peritoneal washings from one *R. exulans* to another was successful. However, meronts ("cysts") were found with a dissecting microscope in the tissues of both rats and mice 30–60 days after the injection of oocysts. These "cysts" could be seen with the naked eye in mice between 40 and 60 days. They had degenerated in this host by 70 days, but persisted in *R. norvegicus* for at least 393 days and in *R. exulans* for 64 days. *R. norvegicus* is thus the best intermediate host.

These meronts ("cysts") consisted of a host cell (probably a fibroblast) forming a thick wall with multiple and hypertrophied nuclei and containing some hundreds or thousands of bradyzoites. The meronts were approximately spherical, up to 200 μm in diameter in sections, with a wall up to 30 μm thick.

After newborn kittens had ingested meronts containing merozoites in mouse tissues, Frenkel (1977) observed one generation of asexual reproduction. Meronts 500–800 μm developed between days 4 and 14 in the lamina propria of the small intestine; development was asynchronous, both immature and mature meronts being present between 10 and 13 days. They generally extended into a blood vessel, appearing to be in endothelial or intimal cells. They were also seen in the liver after 10 days.

Gametogony. Frenkel (1977) found macrogamonts 10-13 μm in diameter in goblet cells in the small intestine of these kittens 13-16 days after ingestion of mouse meronts; they were usually between the vacuole and the nucleus. He found microgamonts up to 11 μm in diameter on day 13. Oocysts were then formed and extruded.

Prepatent Period. 12-15 days (Frenkel, 1977).

Patent Period. 5-12 days (Frenkel, 1977).

Pathogenicity. There is no evidence that this species is pathogenic, either for cats or rodents.

Immunity. Frenkel (1977) found serologic cross-reactions in the indirect fluorescent antibody test between *B. wallacei* and *B. jellisoni, Toxoplasma gondii,* and *Sarcocystis muris;* he found cross-reactions with *B. jellisoni* but not with *T. gondii* in the dye test. Mice previously infected with *B. wallacei* developed immunity to reinfection and to *B. jellisoni.*

Cross-Transmission Studies. Frenkel (1977) could not transmit this species from the cat to *R. rattus* or the golden hamster *Mesocricetus auratus.* Fayer and Frenkel (1979) could not infect calves, as either intermediate, transport, or definitive hosts.

Cultivation. This species has not been cultivated, so far as is known.

Remarks. This organism was isolated by G. D. Wallace from a stray cat on Oahu, Hawaii (Wallace, 1975).

Besnoitia darlingi (Brumpt, 1913) Mandour, 1965

Synonyms. Sarcocystis darlingi Brumpt, 1913; *Besnoitia panamensis* Schneider, 1965; *Besnoitia sauriana* Garnham, 1966(?).

Type Definitive Host. Domestic cat *Felis catus.*

Intermediate Hosts. The following have been found to be naturally infected: Opossum *Didelphis virginiana,* basilisk lizards *Basiliscus basiliscus* and *B. vittatus,* "borriguero" lizard *Ameiva ameiva praesignis.*

Other Intermediate Hosts. The following have been infected experimentally: marmoset *Saguinus geoffroyi,* house mouse *Mus musculus,* golden hamster *Mesocricetus auratus,* free-tailed bat *Carollia perspicillata.*

Location. Gamonts, gametes, zygotes, and oocysts presumably occur in the intestinal wall of the cat. Meronts containing "innumerable" bradyzoites occur in the tongue, ears, mesenteries, gastrointestinal tract, adrenal glands, kidneys, abdominal muscles, and uterus of the opossum; they are most numerous in the adrenal glands (Smith and Frenkel, 1977). Meronts are in the lungs, kidneys, heart, spleen, brain, muscles, fibrous tissue, etc. They are in the peritoneal fluid of experimentally infected laboratory mice and marmosets. Tachyzoites are in the lungs and lymph nodes of experimental mice (Smith and Frenkel, 1977).

Geographic Distribution. Central America (Panama, British Honduras), North America (Illinois, Kansas, Kentucky, Missouri).

Prevalence. Common in opossums. Flatt, Nelson, and Patton (1971) found this species in the tissues of 8 out of 13 opossums *D. virginiana* from Missouri and Illinois. Prevalence in cats unknown.

Oocyst Structure. Unsporulated oocysts in cat feces 11–13 (mean, 12) μm in diameter, with a finely granular sporont almost completely filling the oocyst wall. Sporulated oocysts 11–13 x 10–11 (mean, 12 x 10) μm, without micropyle or residuum. Sporocysts ellipsoidal, 6–9 x 5–6 (mean, 8 x 5) μm, without Stieda body, with residuum. Sporozoites about 5 μm long (Smith and Frenkel, 1977).

Sporulation. 2–3 days at room temperature in sucrose solution (Smith and Frenkel, 1977).

Merogony. The life cycle has not been determined. Schneider (1965) said that the merozoites in peritoneal exudate were crescentic, spindle-shaped, piriform, or ovoid, 7–11 x 2–3 μm. He said that the meronts in the organs were ovoid or spherical and might appear lobed; the wall was 3-layered, the outer layer 2.5–3 μm thick, the middle one containing large nuclei, and the inner one a thin membrane. Garnham (1966) said that the wall was 12 μm thick. Schneider (1965) said that the meronts in the organs were 66–300 x 62–156 μm, with thousands of crescent- to banana-shaped merozoites 7–8 x 1–2 μm. Smith and Frenkel (1977) found small polyzoic meronts, most of which contained "innumerable" bradyzoites, in the tongue, ears, mesenteries, gastrointestinal tract, adrenal glands, kidneys, abdominal muscle, and uterus of 2 of 5 road-killed opossums in Kansas. The concentration of meronts was greatest in the adrenal glands. They were 438–827 (mean, 693) μm in diameter in hematoxylin and eosin sections. In mice fed sulfadiazine after having been inoculated subcutaneously or intraperitoneally with oocysts, they found meronts on day 53 in the gastrointestinal tract (mostly the jejunum), cardiac muscle, brain, pancreas, and lung. These meronts were 48–171 (mean, 106) μm in diameter.

Gametogony. Presumably this occurs in the wall of the intestine of the cat. It has not been described.

Prepatent Period. 10 days in the cat (Smith and Frenkel, 1977).

Patent Period. 10 days (Smith and Frenkel, 1977).

Pathogenicity. This species is presumably not pathogenic for cats. Smith and Frenkel (1977) found that 3 mice fed sporulated oocysts from the cat died in 11–14 days from besnoitiosis. In addition, they noted necrosis, fibrosis, and compensatory hyperplasia in the adrenal glands of some naturally infected opossums.

Immunity. Smith and Frenkel (1977) reported serum titers of 1:384 and 1:1,024 in mice inoculated subcutaneously and intraperitoneally, respectively, with oocysts from the cat. The sera of these 2 mice were negative to *Toxoplasma gondii* antigen in the dye test.

Cross-Transmission Studies. Schneider (1965) was unable to infect the guinea pig, rabbit, a rhesus monkey, a white-faced monkey *Cebus capucinus imitator,* and pigeons with this species. Garnham (1966) was unable to infect white mice. Smith and Frenkel (1977) could not infect the dog, but transmitted this species from the cat to the house mouse via oocysts, from the opossum to the cat, house mouse, and golden hamster via tissue meronts, and from mouse to mouse via tissue meronts. Wallace and Frenkel (1975) did not obtain oocysts in the domestic cat, bobcat, cougar, owl, hawk, boid snake, colubrid snake, or viperid snake (they did not give scientific names).

Besnoitia sp. Ito, Tsunoda, and Shimura, 1978

Synonym. Large type of *Isospora bigemina* of the cat of Ito, Tsunoda, and Shimura (1978).

Type Definitive Host. Domestic cat *Felis catus.*

Intermediate Hosts. The following are all experimental intermediate hosts: Norway rat *Rattus norvegicus,* Mongolian gerbil *Meriones unguiculatus,* house mouse *Mus musculus,* vole *Microtus montebelli,* domestic rabbit *Oryctolagus cuniculus,* golden hamster *Mesocricetus auratus.*

Location. Oocysts in cat intestine, meronts in brain, mesenteric lymph nodes, intestine, and many other organs including the muscles of intermediate hosts.

Geographic Distribution. Asia (Japan).

Prevalence. Unknown; this organism was found in a single cat.

Oocyst Structure. Subspherical or ellipsoidal, 14–18 x 12–15 (mean, 16 x 13) μm, with smooth, colorless wall, without micropyle, residuum, or polar granule; sporocysts elongate ellipsoidal, 10–14 x 8–9 (mean, 13 x 8) μm, without Stieda body, with residuum.

Sporulation. 72–74 hours at 25 C.

Merogony and Gametogony. Ito, Tsunoda, and Shimura (1978) found young meronts ("cysts") in the brain, mesenteric lymph nodes, and intestine of laboratory rats 17 days after oral inoculation. They had developed into mature *Besnoitia*-like "cysts" by 27 days after inoculation. Most were spherical, about 150–250 μm in diameter, and PAS-positive. "Cysts" were widely distributed in the intermediate hosts, being found in the brain, lungs, mesenteric lymph nodes, tongue, esophagus, stomach, intestine, thymus, pancreas, testis, adrenal gland, submaxillary gland, heart, diaphragm, uterus, gluteal and other muscles. They found "cysts" in the intermediate hosts 343 (mouse) or 249 (rat) days after feeding.

In cats, Ito, Tsunoda, and Shimura (1978) found both sexual and asexual stages in the intestinal cells. They found young and giant meronts 50–120 x 35–80 μm in the lamina propria of the jejunum 6–8 days after inoculation, before any other developmental stage was seen. This was the first stage. In the second stage they saw giant meronts in the same location 8–10 days after feeding. They saw liberated merozoites around both types of meront. In the third stage, 12 days or more after feeding, they saw merozoites, giant meronts, rather small, mature meronts, macrogametes, microgamonts, zygotes, and unsporulated oocysts in the intestine.

They postulated that the life cycle was as follows: oocysts are passed in the cat feces. They sporulate on the ground. When eaten by an intermediate host, the sporozoites emerge, enter host cells, and develop to mature "cysts" containing bradyzoites. When a cat eats the intermediate host, the bradyzoites enter the lamina propria of the intestine and turn into meronts that grow to giant size and produce a large number of merozoites. These in turn enter new host cells in the intestine (they did not say where) and turn into meronts, which produce further merozoites that turn into macrogametes or microgamonts. The latter form biflagellate microgametes, fertilization takes place, and unsporulated oocysts are passed. Thus, there is 1 asexual generation in the intermediate host and apparently 2 in the cat.

Prepatent Period. Usually 11–15 (mainly 12–13), and up to 22 days (Ito, Tsunoda, and Shimura, 1978).

Patent Period. Usually 8–20 days, sometimes longer (Ito, Tsunoda, and Shimura, 1978).

Pathogenicity. Ito, Tsunoda, and Shimura (1978) saw no clinical signs or pathologic changes either in the cat or the intermediate hosts.

Cross-Transmission Studies. Ito, Tsunoda, and Shimura (1978) found that *R. norvegicus, M. unguiculatus, M. musculus, M. montebelli, O. cuniculus,* and *M. auratus* could act as intermediate hosts, but not the guinea pig *Cavia porcellus,* dog, cat, or chicken. The Norway rat and Mongolian gerbil were more susceptible than mice, and the golden hamster was by far the least susceptible. They were unable to infect 11 kittens by oral inoculation with oocysts, neither oocysts nor developmental stages being produced. They were unable to transmit the organism from mouse to mouse or rat to rat and concluded that it has an obligatorily heteroxenous life cycle.

Remarks. Ito, Tsunoda, and Shimura (1978) said that there were 2 differences between this form and *B. wallacei* and that more data were needed to determine whether their form was *B. wallacei* or another species. The differences were (1) the meronts in the cat were 50–120 x 35–80 μm in their form, whereas the giant meronts seen in the lamina propria of the cat by Frenkel (1977) were 500–800 μm in diameter when mature; (2) the hamster was slightly susceptible in their study, and not susceptible in that of Frenkel (1977). A third difference is that Frenkel (1977) found only 1 generation of meront in the cat rather than 2.

Host Genus *Leo*

Eimeria anekalensis Rajasekariah et al., 1971

(Plate 9, Fig. 51)

Type Host. Leopard *Leo* (synonym, *Felis*) *pardus.*
Location. Unknown; oocysts found in feces.
Geographic Distribution. Asia (India).
Oocyst Structure. Oocysts ovoid, 22–30 x 18–22 (mean, 26.5 x
20) μm, with smooth, colorless, 2-layered wall, the outer thinner
and darker than the inner, with micropyle and polar granule, without residuum. Sporocysts spindle-shaped, 10–15 x 5–8 (mean, 12
x 7) μm, with Stieda body, without residuum. Sporozoites 7–12 x
3–6 (mean, 10 x 4) μm.
Sporulation. 40–42 hours in 2.5% potassium bichromate solution.
Remarks. Rajasekariah et al. (1971) found this species in a "panther" cub kept at the Dharmaram College, Bangalore. Whether it is
a true parasite of the leopard remains to be determined.

Eimeria novowenyoni Rastegaieff, 1929

(Plate 9, Fig. 53)

Type Definitive Host. Tiger *Leo* (synonym, *Felis*) *tigris.*
Other Definitive Host. Leopard *Leo* (synonym, *Felis*) *pardus* (?).
Location. Unknown; oocysts found in feces.
Geographic Distribution. Tiger, USSR (Leningrad zoo). Leopard,
Asia (India–captured "panther" cub kept at Dharmaram College,
Bangalore).
Oocyst Structure. Oocysts spherical, 14–15 μm in diameter,
without micropyle. Rastegaieff (1929, 1930), who named this
species from the tiger, gave no further structural information ex-
cept to say that it formed 4 sporocysts after sporulation.
Rajesekariah et al. (1971), who found it in a captured "panther"
cub, said that its oocysts were almost spherical, 18–20 μm in dia-

meter, with a 2-layered wall of which the outer was 0.5 μm thick and the inner 1.5 μm thick, without a micropyle, residuum, or polar granule, that its sporocysts were ellipsoidal, 10 x 6 μm, without a Stieda body or residuum, and that its sporozoites were 8 x 4 μm, with one end rounded and the other pointed.

Sporulation. 40–42 hours (Rajasekariah et al., 1971).

Remarks. Whether this is a true parasite of the tiger remains to be determined, as does whether it occurs in both the tiger and leopard.

Eimeria (?) *hartmanni* Rastegaieff, 1930

(Plate 12, Fig. 68)

Type Definitive Host. Tiger *Leo* (synonym, *Panthera*) *tigris.*
Other Definitive Host. Leopard *Leo* (synonym, *Felis*) *pardus* (?).
Location. Unknown; oocysts found in the feces.
Geographic Distribution. Tiger, USSR (Leningrad zoo). Leopard, Asia (India—captured "panther" cub kept at Dharmaram College, Bangalore).
Oocyst Structure. Oocysts ovoid, 23 x 14 μm, with a markedly flattened small end bearing a broad micropyle. Rastegaieff (1930) gave no further information; the oocysts did not sporulate. Rajasekariah et al. (1971) found what they assigned to this species in a leopard ("panther") cub captured in India and kept at Dharmaram College, Bangalore. Its oocysts were elongated, ovoid to ellipsoidal, with a flattened end, 20–23 x 14–19 (mean, 21.5 x 16.5) μm with an inconspicuous micropyle, without a residuum, or polar granule, with a 2-layered wall, and with spindle-shaped sporocysts 9–10 x 6–8 μm, with a Stieda body, without a residuum.
Sporulation. 40–42 hours (Rajasekariah et al. (1971).
Remarks. Rastegaieff's form sounds suspiciously like a rabbit coccidium. Whether this is a true parasite of the tiger remains to be determined, as does whether it occurs in both the tiger and leopard.

Isospora leonina Mandal and Ray, 1960

Type Definitive Host. Lion *Leo* (synonym, *Felis*) *leo.*

Location. Unknown; oocysts found in feces.

Geographic Distribution. Asia (India—Calcutta zoo), Africa (?).

Oocyst Structure. Oocysts said to be rhomboidal but illustrated as essentially ellipsoidal, 30–32 x 28–31 (mean, 32 x 28) μm, with 2-layered wall, outer layer colorless, 1 μm thick, inner layer light orange, 1.5 μm thick, without micropyle or residuum. Sporocysts 16–20 x 13–15 μm, with residuum. Sporozoites sausage-shaped. No other information given.

Patnaik and Acharjyo (1970), who reported this species from a lion in Africa, said that it was 30–33 (mean, 32 x 29) μm, with a thin wall, without micropyle, and that the sporonts were spherical, 22–23 μm in diameter.

Sporulation. 1–2 days at 33 C in 2.5% potassium bichromate solution.

Cross-Transmission Studies. Mandal and Ray (1960) failed to infect a puppy with this species.

Remarks. Whether this is a true parasite of the lion remains to be determined.

Isospora sp. Pande et al. 1970

Type Definitive Host. Lion *Leo* (synonym, *Panthera*) *leo*.

Location. Unknown; oocysts found in feces.

Geographic Distribution. Asia (India—Mathura zoo).

Oocyst Structure. Oocysts ovoid, 23–33 x 20–28 (mean, 26 x 22) μm, with 2-layered wall 1.3 μm thick, outer layer straw-colored, inner layer dark brown, without micropyle, residuum, or polar granule. Sporocysts ellipsoidal, 16–18 x 11–13 (mean, 17 x 12) μm, with 1-layered wall, without Stieda body, with residuum. Sporozoites banana-shaped, 12–15 x 3–9 (mean, 13 x 4) μm, with one end broad and the other pointed, with a central nucleus and a prominent clear globule at the broad end.

Remarks. Whether this is a true parasite of the lion remains to be determined.

Sarcocystis sp. Bhatvadekar and Purchit, 1963

This organism was found in the heart muscles of 2 lions *Leo leo* that died in a zoo in Bombay, India. The sarcocysts had a very thin hyaline capsule, no septa, and contained numerous crescentic merozoites about 12 x 2 μm. The lions were probably aberrant hosts, the sarcocysts occurring normally in some other animal.

Discussion

As stated in the Introduction, there are 7 families, 101 genera, and about 248 species in the mammalian order Carnivora. Coccidia have been found in hosts belonging to all the families, but in far from all genera or species. Including experimental hosts, coccidia have been found in the following:

Family CANIDAE
    4 (29%) of 14 genera
    11 (about 31%) of about 35 species
Family URSIDAE
    2 (25%) of 8 genera
    2 (about 20%) of about 10 species
Family PROCYONIDAE
    3 (38%) of 8 genera
    3 (about 17%) of about 18 species
Family MUSTELIDAE
    7 (28%) of 25 genera
    12 (about 17%) of about 70 species
Family VIVERRIDAE
    5 (14%) of 37 genera
    7 (about 9%) of about 75 species
Family HYAENIDAE
    1 (33%) of 3 genera
    1 (25%) of 4 species
Family FELIDAE
    3 (50%) of 6 genera
    14 (39%) of 36 species

TOTAL
25 (25%) of 101 genera
50 (about 20%) of about 248 species

This is a better coverage than that reported by Levine and Ivens (1965b) from the 2,688 species of rodents (8.4%) and about the same as that reported by Levine and Ivens (1970) from the 188 species of ruminants (21%). However, it should be pointed out that this figure includes experimental hosts, that it includes quite a number of dubious species of parasites, and that only the domestic dog and cat have been studied at all well. The great majority of descriptions of coccidia have been more or less casual, and complete life cycles are known for very few. It has been only recently, too, that it was realized that *Toxoplasma, Besnoitia, Sarcocystis,* and *Frenkelia* were coccidia and that some coccidian life cycles are heteroxenous. Gaps in our knowledge are obvious to anyone who reads this monograph, and we hope that their existence will stimulate some of our readers to fill them.

Not many years ago it was believed that dogs and cats had the same species of coccidia and that these were 3 in number; the largest species was *Isospora felis,* the smallest *I. bigemina,* and the middle one *I. rivolta.* However, cross-transmission studies and the recognition that *Toxoplasma* and *Sarcocystis* both form oocysts in the intestine of these animals have changed these views. At present, it is known that the coccidia of the dog and cat are all different and that there are at least 15 validly named species in the dog and at least 15 others in the cat. There are undoubtedly more. Those that we accept in the dog are *I. bahiensis, I. burrowsi, I. canis, I. neorivolta, I. ohioensis, I. vulpina, Sarcocystis bertrami, S. bigemina, S. cruzi, S. equicanis, S. fayeri, S. hemionilatrantis, S. levinei, S. miescheriana,* and *S. tenella.* In the cat, we accept *I. felis, I. rivolta, S. cuniculi, S. fusiformis, S. gigantea, S. hirsuta, S. leporum, S. muris, S. porcifelis, Toxoplasma gondii, T. hammondi, T. pardalis, Besnoitia besnoiti, B. darlingi,* and *B. wallacei.*

Sarcocystis species pass sporulated sporocysts in the feces, while the other genera pass unsporulated oocysts. Most species of *Sarcocystis* cannot be differentiated on structural grounds, so far as is known, and the same is true of *Toxoplasma* and of *B. besnoiti* and *B. wallacei.* There are actually 5 structurally recognizable forms in the dog and 5 in the cat. If there are additional structural differences between the various species, only future research will reveal it.

The present monograph contains accounts of 102 named species of coccidia: 39 of *Eimeria*, 39 of *Isospora*, 18 of *Sarcocystis*, 3 of *Toxoplasma*, and 3 of *Besnoitia*. Future research on these same species may decrease the number of *Eimeria* species and dictate shifts in the genera of some of the others.

The great majority of coccidia of prey animals belong to the genus *Eimeria*, while *Isospora* is found with greater frequency among predators. Levine and Ivens (1965b), for instance, reported accounts of 204 species of *Eimeria* and only 10 of *Isospora* in rodent hosts, and Levine and Ivens (1970) reported accounts of 97 species of *Eimeria* and only 4 of *Isospora* in ruminant hosts. In carnivores, in contrast, we have found accounts of about equal numbers of *Eimeria* and *Isospora* species, and it is likely that some of the former are actually parasites of animals that the predators have eaten. Further, the discovery that some coccidia formerly believed to be *Isospora* are actually *Toxoplasma, Besnoitia,* or *Sarcocystis,* and the concomitant discovery that there are more species of *Sarcocystis* than formerly believed, are just beginning to have their effects. Becker (1953) remarked on the predator-prey host orientation of *Isospora,* but at that time the life cycles of *Sarcocystis* and *Toxoplasma* were unknown, so the explanation eluded him.

The complete explanation eludes us also. The predator-prey relationship may explain why *Isospora* is relatively common in carnivores. However, it does not explain why it predominates over *Eimeria* in noncarnivores like the passerine birds.

In any study of the coccidia of predators, one must be sure that the coccidia found are not pseudoparasites of the animal in which they were found, and thus true parasites of some other animal that it had eaten. This is not certain in some cases (see *Eimeria canis,* for example). We have indicated the species of which we are suspicious. In order to eliminate the possibility that a coccidium found in a predator is truly a parasite of that predator, it should be known with certainty that it has not eaten another animal for at least 2 days. It is impossible to know this for any wild animal, and even domestic ones or those in zoos may have captured and eaten a mouse or some other rodent.

It seems likely that certain coccidian species are limited to certain geographic localities. This is obvious, of course, for species occurring in definitive hosts whose geographic range is limited, such as the hyena and kinkajou, but it is not so obvious for parasites of domestic carnivores, such as the dog and cat. However, if a suitable host for a heteroxenous coccidium does not exist in some region, one would not expect to find that coccidium there. So far as we are aware, the sporocyst of *S. levinei* look like those of *S. cruzi, S. ovicanis, S. miescheriana, S. bertrami,* and *S. hemionilatrantis,* but one would not expect to find it in the United States, since its intermediate host is the water buffalo, which is rare outside of zoos in the United States. Similarly, one would not expect to find *S. hemionilatrantis* in the dog outside North America (and only in certain parts thereof), since neither its intermediate host (the mule deer) nor its natural definitive host (the coyote) occurs outside that continent.

In addition, it is likely that the dog or cat might be merely an experimental definitive host and not the normal one for a *Sarcocystis* of some prey animals. This is probably the case for *Sarcocystis* spp. of gazelles and antelope in Africa, as well as for such species as *S. hemionilatrantis* in North America.

A further point should be made. Early describers tended to be extremely sketchy in their descriptions, and this situation is reflected in the descriptions of the coccidia given here. In such cases, new studies and complete descriptions are needed. In this connection, the term "double-contoured," which was once often used, is of little value. We do not know how to interpret it and suspect that it meant different things to different people. It might mean that the oocyst wall is composed of 2 layers, or it might mean that it is composed of 1 layer whose outer and inner edges look heavy. The term should not be used. Another term that should be avoided is oval. It means egg-shaped, with one end smaller than the other, but many authors have thought that it means ellipsoidal, with both ends the same. The term ovoid is preferable because of this confusion in meaning.

Summary

This monograph summarizes the known information on taxonomy, structure, life cycle, hosts, location in the host, geographic distribution, prevalence, sporulation, merogony, gametogony, prepatent period, patent period, pathogenicity, immunity, cross-transmission studies, and cultivation of the 102 named species of coccidia of carnivores. These include 39 species of *Eimeria*, 39 of *Isospora*, 18 of *Sarcocystis*, 3 of *Toxoplasma*, and 3 of *Besnoitia*. In addition, similar data are given for those forms for which insufficient information is available to justify assigning them names. Modern diagnoses are given for the apicomplexan protozoan taxa, which contain species of coccidia in the Carnivora. The confused taxonomy of the coccidia of carnivores is corrected.

Species of coccidia have been named from 25% of the 101 genera and about 20% of the approximately 248 species of carnivores. Experimental hosts are included.

The following new species are named: *Eimeria voronezhensis* n. sp. for *E. mephitidis* Andrews, 1928 of Yakimoff and Matikaschwili (1932) from the striped skunk *Mephitis mephitis; Isospora chobotari* n. sp. for *Isospora* sp. Inabnit, Chobotar, and Ernst, 1972 from the raccoon *Procyon lotor; Isospora lyncis* n. sp. for *I. felis* (Wasielewski, 1904) Wenyon, 1923 of Triffitt (1927) from the lynx *Lynx* sp. The following new combinations are introduced: *Isospora* (?) sp. for *Hammondia* sp. Ashford, 1977; *Toxoplasma pardalis* for *Hammondia pardalis* Hendricks et al., 1979.

Glossary

Apical Complex. The anterior structures of certain cells in the phylum Apicomplexa, including the polar ring(s), conoid, rhoptries, micronemes, and subpellicular microtubules.

Bradyzoite. A slowly developing merozoite. This term is used especially for the merozoites in the last-generation meronts (sarcocysts) of *Sarcocystis* and for the merozoites in pseudocysts of *Toxoplasma*.

Conoid. An electron-dense, hollow structure in the form of a truncated cone inside the polar ring(s) at the anterior end of certain stages of most Apicomplexa; composed of spirally coiled microtubules.

Cyst. A resistant stage of an organism formed by the organism's laying down a wall around itself.

Endodyocyte. A trophozoite formed by endodyogeny.

Endodyogeny. Formation of 2 daughter individuals by internal budding.

Endogenous Cycle. That part of the life cycle that takes place within a host.

Endopolygeny. Formation of more than 2 daughter cells (usually merozoites) by internal budding.

Euryxenous. Having a broad host range (considered as occurring in more than 1 host order). Most avian malaria parasites are euryxenous. The asexual part of the life cycle of *Toxoplasma* is euryxenous.

Exogenous Cycle. That part of the life cycle that takes place outside the host.

Gametocyte. A gamont (i.e., a cell that will form gametes).

Gametogony. Formation of gametes, often by schizogony.

Gamont. An individual that will produce 1 or more gametes. In the Apicomplexa the gamonts are already haploid, so far as is known.

Globule, Clear. A body in some coccidian sporozoites, apparently composed of protein plus amylopectin, that stains with eosin and has a homogeneous, clear structure. Often called a refractile body.

Globule, Eosinophilic. A clear globule.

Heteroxenous. Having 2 or more types of host in the life cycle.

Homoxenous. Having 1 type of host in the life cycle.

Host. The organism that harbors a parasite.

Hypnozoite. A nonmultiplying sporozoite in a transport host.

Macrogamete. A relatively large gamete, considered female.

Macrogametocyte. A macrogamont.

Macrogamont. A gamont that will turn into a macrogamete. So far as is known, all coccidian macrogamonts are already haploid. (In the coccidia macrogamonts are simply young macrogametes.)

Merogony. Formation of merozoites.

Meront. An asexual stage in the life cycle that divides by schizogony (multiple fission) to form merozoites.

Merozoite. A cell produced by merogony in the asexual part of an apicomplexan life cycle. It forms either a new meront (schizont) or a gamont.

Mesoxenous. Having a moderate host range (considered as occurring in more than 1 host family within a host order).

Metrocyte. A "mother cell." A premerozoite found in the last-generation meronts of *Sarcocystis* and *Frenkelia*. It is rather stout and ellipsoidal, with a deeply folded cell surface and few, if any, elements of an apical complex. It divides repeatedly by endodyogeny to form fresh metrocytes, which become progressively more elongate and eventually become typical merozoites with a complete apical complex.

Microgamete. A relatively small gamete, considered male.

Microgametocyte. A microgamont.

Microgamont. A gamont that will produce microgametes, the number depending on the parasite group.

Microneme. An elongate, electron-dense organelle extending longitudinally in the anterior part of the body of certain stages in certain Apicomplexa, possibly attached to and giving rise to the rhoptries.

Micropore. An opening in the body visible only with the electron microscope through which particulate food or other material can be taken into the body. Sometimes called a cytostome, micropyle, or cuticular pore.

Microtubule, Subpellicular. A slender, electron-dense, hollow structure extending back from the polar ring region just beneath the pellicle, of the sporozoite, merozoite, and sometimes other stages of the Apicomplexa. Visible only with the electron microscope.

Micropyle. An opening (or the position of an opening) in the wall of an oocyst.

Monoxenous Parasite. (1) A parasite with a single species of host in its life cycle. (2) A parasite with 1 type of host in its life cycle. (This definition is often used, but "homoxenous" is preferable.)

Oocyst. A cyst formed around a zygote.

Plastic Granule. A granule in coccidian macrogametes that will later participate in forming the oocyst wall. A wall-forming body.

Polar Granule. A structure found in the oocysts of some coccidia, formed by the first (reduction) division of the zygote in the oocyst.

Polar Ring. An electron-dense ring at the anterior end of certain stages of the Apicomplexa. Visible only with the electron microscope.

Refractile Body. A clear globule present in some coccidian sporozoites, apparently composed of protein plus amylopectin. "Clear globule" is preferable.

Residuum, Oocyst. The material, exclusive of the polar granule(s), remaining in an oocyst after formation of sporocysts or sporozoites.

Residuum, Sporocyst. The material remaining in a sporocyst after formation of sporozoites.

Rhoptry. The electron-dense, tubular or saccular organelle, often enlarged at the posterior end, extending back from the anterior region in the sporozoite, merozoite, and sometimes other stages of the Apicomplexa. Superseded synonyms: toxoneme, paired organelle, lankersterelloneme, eimerianeme, dense body.

Sarcocyst. The last-generation meront of *Sarcocystis* in the muscles of the intermediate host.

Schizogony. Formation of daughter cells by multiple fission. If the daughter cells are merozoites, the schizogony may be called merogony. If they are sporozoites, the schizogony may be called sporogony. If they are gametes, the schizogony may be called gametogony. (Some people limit the term "schizogony" to merogony.)

Schizont. A stage in the life cycle that divides by schizogony (multiple fission). This name is often used only for the asexual stage that produces merozoites, i.e., for a meront.

Spore. A resistant stage formed within a cell. In the Apicomplexa this term is often applied to the sporocysts of coccidia or oocysts of gregarines, and in this case apparently signifies a walled body containing or producing one or more uninucleate bodies (sporozoites) capable of developing into other stages in the life cycle; this usage is erroneous.

Sporocyst. A cyst formed within the oocysts of most coccidia; it contains the sporozoites. (The term sporocyst is often applied to the oocyst of gregarines, but these protozoa actually have no sporocyst.)

Sporogony. Formation of sporozoites by division of a zygote, often by a type of schizogony.

Sporont. The stage in the life cycle of coccidia that will form sporocysts, i.e., the zygote within the oocyst wall.

Sporozoite. An infective stage produced by sporogony, usually with an envelope or shell.

Sporulation. The process of sporozoite formation, i.e., sporogony.

Stenoxenous. Having a narrow host range (considered as occurring in a single host family). Most coccidia and mammalian malaria parasites are stenoxenous.

Stieda Body. A knoblike thickening at one end of the sporocyst wall in coccidia.

Substiedal Body. A dependent body attached to the Stieda body and extending into the sporocyst.

Tachyzoite. A fast-developing merozoite. Term used especially in referring to the merozoites found in the group stages (aggregates)

of *Toxoplasma* in the acute stage of the infection or in the early
generation(s) of *Sarcocystis* in endothelial or other cells.

Vacuole, Parasitophorous. The vacuole in a host cell that contains
a stage of an apicomplexan parasite.

Wall-Forming Body. A body in a coccidian macrogamete that will
later participate in formation of the oocyst wall. A plastic granule.

Literature Cited

Adler, S. 1924. An *Isospora* of civet cats. Ann. Trop. Med. Parasit. 18:87-94.

Agostinucci, G. and E. Bronzini. 1955 (1953). *Eimeria genettae* n. sp. parasita della *Genetta dongolana*. Proc. 6th Internatl. Congr. Microbiol. 5:284-85.

Akinchina, G.T. and J.M. Doby. 1969. Étude comparée de la multiplication de *Toxoplasma gondii* et de *Besnoitia jellisoni* dans les cultures de cellules. Prog. Protozool. 3:222.

Alcaino, H. and I. Tagle, 1970. Estudio sobre enteroparasitosis del perro en Santiago. Bol. Chil. Parasit. 25:5-8.

Alwar, V.S. and C.M. Lalitha. 1958. Parasites of domestic cats *(Felis catus)* in Madras. Ind. Vet. J. 35:292-95.

De Amaral, V., R.C. Amaro, and E.H. Birgel. 1964. Sobre a presença de *Isospora felis* Wenyon, 1923 (Protozoa: Eimeriidae Poche, 1913) em *Canis familiaris*, em São Paulo. Arq. Inst. Biol. S. Paulo 31:101-2.

Andrews, J.M. 1926. The specificity of the *Isospora* of cats and dogs with respect to host. J. Parasit. 13:89.

Andrews, J.M. 1928. New species of coccidia from the skunk and prairie dog. J. Parasit. 14:193-94.

Anpilogova, N.V. and A.I. Sokov. 1973. Novye vidy koktsidii iz blednoi ili sredneaziatskoi, rysi v Tadzhikistane. Izvest. Akad. Nauk Tadzhik. SSR, Biol. Nauk 1973 (4):89-90.

Arther, R.G. and G. Post. 1977. Coccidia of coyotes in eastern Colorado. J. Wildl. Dis. 13:97-100.

Aryeetey, M.E. and G. Piekarski. 1976. Serologische *Sarcocystis*-Studien am Menschen und Ratten. Z. Parasitenk. 50:109-24.

Ashford, R.W. 1977. The fox, *Vulpes vulpes*, as a final host for Sarcocystis of sheep. Ann. Trop. Med. Parasit. 71:29–34.

Ashford, R.W. 1978. *Sarcocystis cymruensis* n. sp., a parasite of rats *Rattus norvegicus* and cats *Felis catus*. Ann. Trop. Med. Parasit. 72:37–43.

Atkinson, E.M. 1978. A *Sarcocystis* species of rats. J. Protozool. 25:13B.

Babudieri, B. 1932. I sarcosporidi e le sarcosporidiosi (Studio monografico). Arch. Protistenk. 76:421–580.

Balfour, A. 1913. A sarcocyst of a gazelle *(G. rufifrons)* showing differentiation of spores by vital staining. Parasitology 6:52–56.

Balozet, L. 1933. Sur une coccidie de la mangouste. Bull. Soc. Path. Exot. 26:913–14.

Barreto, J.F. and J.L. de Almeida. 1937. Primeiras observacões sobre a presenca de *Isospora felis* Wenyon, 1923 (Protozoa-Eimeridia) em felideos no Brasil. Bol. Soc. Bras. Med. Vet. 7:357–60.

Barriga, O.O. and S. Jaramillo. 1966. Encuesta enteroparasitaria en perros de Castro, Provincia de Chiloe (Chile). Rev. Soc. Med. Vet. (Chile) 16:9–16.

Basson, P.A., R.M. McCully, and R.D. Bigalke. 1970. Observations on the pathogenesis of bovine, and antelope strains of *Besnoitia besnoiti* (Marotel, 1912) infection in cattle and rabbits. Onderstepoort J. Vet. Res. 37:105–26.

Bearup, A.J. 1954. The coccidia of carnivores in Sydney. Austral. Vet. J. 30:185–86.

Bearup, A.J. 1960. Parasitic infection in cats in Sydney, with special reference to the occurrence of *Ollulanus tricuspis*. Austral. Vet. J. 36:352–54.

Becker, B., H. Mehlhorn, and A.-O. Heydorn. 1979. Light and electron microscopic studies on gamogony and sporogony of 5 *Sarcocystis* species in vivo and in tissue cultures. Zbl. Bakt. Hyg., I. Abt. Orig. A 244:394–404.

Becker, E.R. 1934. A check-list of coccidia of the genus *Isospora*. J. Parasit. 20:195–96.

Becker, E.R. 1953. How parasites tolerate their hosts. J. Parasit. 39:467–80.

Bernard, P.N. and J. Bauche. 1912. Filariose et atherome aortique du buffle et du boeuf. Bull. Soc. Path. Exot. 5:109–14.

Beverley, J.K.A. 1973. Vertical transmission of *Toxoplasma gondii.* Prog. Protozool. 4:41.

Beverley, J.K.A. 1974. Some aspects of toxoplasmosis, a worldwide zoonosis. *In* Soulsby, E.J.L., ed. *Parasitic Zoonoses. Clinical and Experimental Studies.* Academic Press, New York, pp. 1-25.

Beverley, J.K.A. 1976. Toxoplasmosis in animals. Vet. Rec. 99: 123-27.

Bhatavdekar, M.Y. and B.L. Purchit. 1963. A record of sarcosporidiosis in lion. Ind. Vet. J. 40:44-45.

Bigalke, R.D. 1962. Preliminary communication on the cultivation of *Besnoitia besnoiti* (Marotel, 1912) in the tissue culture and embryonated eggs. J. S. Afr. Vet. Med. Assoc. 33:523-32.

Bigalke, R.D. 1968. New concepts on the epidemiological features of bovine besnoitiosis as determined by laboratory and field investigations. Onderstepoort J. Vet. Res. 35:3-138.

Bigalke, R.D., P.A. Basson, R.M. McCully, P.P. Bosman, and J.H. Schoeman. 1973. Studies in cattle on the development of a live vaccine against bovine besnoitiosis. Onderstepoort J. Vet. Res. 40:207-9.

Bigalke, R.D. and T.W. Naudé. 1962. The diagnostic value of cysts in the scleral conjunctiva in bovine besnoitiosis. J. S. Afr. Vet. Med. Assoc. 33:21-27.

Bigalke, R.D. and J.H. Schoeman. 1967. An outbreak of bovine besnoitiosis in the Orange Free State, Republic of South Africa. J. S. Afr. Vet. Med. Assoc. 38:435-37.

Bigalke, R.D., J. Schoeman, and R. McCully. 1974. Immunizations against bovine besnoitiosis with a live vaccine prepared from a blue wildebeest strain of *Besnoitia besnoiti* grown in cell cultures: I. Studies on rabbits. Onderstepoort J. Vet. Res. 41:1-6.

Bigalke, R.D. and R.C. Tustin. 1960. The occurrence of a cyst of *Sarcocystis* Lankester 1882 in the cerebellum of an ox. J. S. Afr. Vet. Med. Assoc. 31:271-74.

Biocca, E. 1957 (1956). Alcune considerazioni sulla sistematica dei protozoi e sulla utilita di creare una nuova classe di protozoi. Rev. Brasil. Malariol. 8:91-102.

Biocca, E. 1968. Class Toxoplasmatea: Critical review and proposal of the new name *Frenkelia* gen. n. for M-organism. Parassitologia 10:89-98.

Biocca, E., T. Balbo, E. Guarda, and R. Costantini. 1975. L'importanza della volpe *(Vulpes vulpes)* nella trasmissione della sarcosporidiosi dello stambecco *(Capra ibex)* nel Parco Nazionale del Gran Paradiso. Parassitologia 17:17-24.

Blanchard, R. 1885. Note sur les sarcosporidies et sur un essai de classification de ces sporozoaires. Bull. Soc. Zool. 10:244-77.

Blažek, K., A. Kotrlý, and R. Ippen. 1976. Sarkosporidioza myokardu sparkate zvere. Vet. Med., Praha 49:75-80.

Bledsoe, B. 1976. *Isospora vulpina* Nieschulz and Bos, 1933; Description and transmission from the fox *(Vulpes vulpes)* to the dog. J. Protozool. 23:365-67.

Bledsoe, B. 1976a. Transmission of *Isospora vulpina* from the silver fox to the dog; establishment of the mouse as an intermediate host of *I. vulpina.* Prog. Abstr. An. Mtg. Am. Soc. Parasit. 51:46.

Boch, J., K.E. Laupheimer, and M. Erber. 1978. Drei Sarkosporidienarten bei Schlachtrindern in Suddeutschland. Berl. Münch. Tierärztl. Wochenschr. 91:426-31.

Boch, J., U. Mannewitz, and M. Erber. 1978. Sarkosporidien bei Schlachtschweinen in Suddeutschland. Berl. Münch. Tierärztl. Wochenschr. 91:106-11.

Bray, R.S. 1954. On the coccidia of the mongoose. Ann. Trop. Med. Parasit. 48:405-15.

Brumpt, E. 1913. *Précis de Parasitologie.* 2nd ed. Masson, Paris. 1011 p.

Burrows, R.B. 1968. Internal parasites of dogs and cats from central New Jersey. Bull. N. J. Acad. Sci. 13(2):3-8.

Burrows, R.B. and G.R. Hunt. 1970. Intestinal protozoan infections in cats. J. Am. Vet. Med. Assoc. 157:2065-67.

Bwangamoi, O. 1968. The incidence of skin diseases of cattle in Uganda. Bull. Epizoot. Dis. Afr. 16:115-19.

Carini, A. and F. da Fonseca. 1938. Sobre uma nova *Eimeria (E. irara* n. sp.) parasita da *Tayra barbara.* Arq. Biol. S. Paulo 22:36.

Carini, A. and D. Grechi. 1938. Sobre uma nova *Eimeria,* parasita do Nasua nasica. Arq. Biol. S. Paulo 22:104-5.

Carini, A. and L. Migliano. 1916. Sobre um Toxoplasma da cobaya *(Toxoplasma caviae).* Ann. Paulist. Med. Cirurg. 6:113-14.

Castellani, A. 1913. Note on certain cell inclusions. J. Trop. Med. Hyg. 15:354-58.

Catcott, E.J. 1946. The incidence of intestinal protozoa in the dog. J. Am. Vet. Med. Assoc. 108:34-36.

Chessum, B.S. 1972. Reactivation of *Toxoplasma* oocyst production in the cat by infection with *Isospora felis*. Brit. Vet. J. 128: xxiii–xxxvi.

Choquette, L.P.E. and L.G. Gelinas. 1950. The incidence of intestinal nematodes and protozoa of dogs in the Montreal district. Can. J. Comp. Med. Vet. Sci. 14(2):33–38.

Christie, E., J.P. Dubey, and P.W. Pappas. 1976. Prevalence of *Sarcocystis* infection and other intestinal parasitisms in cats from a humane shelter in Ohio. J. Am. Vet. Med. Assoc. 168:421–22.

Clegg, F.G., J.K.A. Beverley, and L.M. Markson. 1978. Clinical disease in cattle in England resembling Dalmeny disease associated with suspected *Sarcocystis* infection. J. Comp. Path. 88: 105–14.

Coles, A.C. 1914. Blood parasites found in mammals, birds, and fishes in England. Parasitology 7:17–61.

Corner, A.H., D. Mitchell, E.B. Meads, and P.A. Taylor. 1963. Dalmeny disease. An infection of cattle presumed to be caused by an unidentified protozoon. Can. Vet. J. 4:252–64.

Costa, H.M.A. and M.G. Freitas. 1959. *Isospora felis* Wenyon, 1923 e *Isospora rivolta* Gassi, 1879, em cães de Belo Horizonte. Arq. Esc. Sup. Vet. 12:127–30.

Coutelen, F.R. 1932. Existence d'une toxoplasmose spontanée et généralisée chez le furet. Un toxoplasme nouveau, *Toxoplasma laidlawi*, n. sp., parasite de *Mustela (Putorius) putorius* var. *furo*. C.R. Soc. Biol. 111:284–87.

Crawley, H. 1914. Two new Sarcosporidia. Proc. Acad. Nat. Sci. Phila. 66:214–18.

Crum, J.M., V.F. Nettles, and W.R. Davidson. 1978. Studies on endoparasites of the black bear *(Ursus americanus)* in the southeastern United States. J. Wildl. Dis. 14:177–86.

Crum, J.M. and A. K. Prestwood. 1977. Transmission of *Sarcocystis leporum* from a cottontail rabbit to domestic cats. J. Wildl. Dis. 13:174–75.

da Cruz, A.A., L. de Sousa, and A. Cabral. 1952. Indice parasitario do *Felis (Felis) catus domesticus* di ciudade de Lisboa. Rev. Med. Vet., Lisbon 47:142–52.

Davalos, A. and C. Briseño. 1960. Coccidiosis por *Isospora felis* en gatos domesticos de la ciudad de Mexico. Rev. Inst. Salub. Enferm. Trop. 20:98–101.

Davis, C.L., T.L. Chow, and J.R. Gorham. 1953. Hepatic coccidiosis in mink. Vet. Med. 48:371–73.

Destombes, P. 1957. Les sarcosporidioses au Vietnam. Bull. Soc. Path. Exot. 50:221–25.

Dissanaike, A.S. and S.P. Kan. 1978. Studies on *Sarcocystis* in Malaysia. 1. *Sarcocystis levinei* n. sp. from the water buffalo *Bubalus bubalis*. Z. Parasitenk. 55:127–38.

Dissanaike, A.S., S.O. Kan, A. Retnasabapathy, and G. Baskavar. 1977. Developmental stages of *Sarcocystis fusiformis* (Railliet, 1897) and *Sarcocystis* sp., of the water buffalo, in the small intestines of cats and dogs, respectively. Southeast Asian J. Trop. Med. Publ. Hlth. 8:417.

Dobell, C.C. 1919. A revision of the coccidia parasitic in man. Parasitology 11:147–97.

Dobell, C.C. and P.W. O'Connor. 1921. *The Intestinal Protozoa of Man.* London. 223 p.

Doflein, F. 1901. *Die Protozoen als Parasiten und Krankheitserreger.* Jena. 287 p.

Dogel', V.A. 1916. Dva novykh' vida *Sarcocystis* iz' afrikanskikh' antilop'. [Two new species of *Sarcocystis* from African antelopes.] Nach. Resul't. Zool. Eksped. (V.A. Dogiel i I.I. Sokolov) Brit. Vost. Africu i Ugandu, 1914. 1 (Art. 8):13 p.

Doran, D.J. 1973. Cultivation of coccidia in avian embryos and cell culture. In Hammond, D.M. with P.L. Long, eds. *The Coccidia.* University Park Press, Baltimore, Md., pp. 183–252.

Dubey, J.P. 1963. Observations on the coccidian oocysts from Indian fox *(Vulpes bengalensis).* Ind. J. Microbiol. 3:143–46.

Dubey, J.P. 1963a. Observations on coccidian oocysts from Indian Hyaena *(Hyaena striata).* Ind. J. Microbiol. 3:121–22.

Dubey, J.P. 1973. Feline toxoplasmosis and coccidiosis. A survey of domiciled and stray cats. J. Am. Vet. Med. Assoc. 162:873–77.

Dubey, J.P. 1975. Experimental *Isospora canis* and *Isospora felis* infection in mice, cats, and dogs. J. Protozool. 22:416–17.

Dubey, J.P. 1975a. *Isospora ohioensis* sp. n. proposed for *I. rivolta* of the dog. J. Parasit. 61:462–65.

Dubey, J.P. 1975b. Experimental *Hammondia hammondi* infection in dogs. Brit. Vet. J. 131:741–43.

Dubey, J.P. 1976. A review of *Sarcocystis* of domestic animals and of other coccidia of cats and dogs. J. Am. Vet. Med. Assoc. 169:1061–1078.

Dubey, J.P. 1977. Attempted transmission of feline coccidia from chronically infected queens to their kittens. J. Am. Vet. Med. Assoc. 170:541–43.

Dubey, J.P. 1977a. Persistence of *Toxoplasma gondii* in the tissues of chronically infected cats. J. Parasit. 63:156–57.

Dubey, J.P. 1978. Pathogenicity of *Isospora ohioensis* infection in dogs. J. Am. Vet. Med. Assoc. 173:192–97.

Dubey, J.P. 1978a. Life-cycle of *Isospora ohioensis* in dogs. Parasitology 77:1–11.

Dubey, J.P. 1979. Life cycle of *Isospora rivolta* (Grassi, 1879) in cats and mice. J. Protozool. 26:433–43.

Dubey, J.P. and R. Fayer. 1976. Development of *Isospora bigemina* in dogs and other mammals. Parasitology 73:371–80.

Dubey, J.P., R. Fayer, and F.M. Seesee. 1978. *Sarcocystis* in feces of coyotes from Montana: Prevalence and experimental transmission to sheep and cattle. J. Am. Vet. Med. Assoc. 173:1167–70.

Dubey, J.P. and J. K. Frenkel. 1972. Extraintestinal stages of *Isospora felis* and *I. rivolta* (Protozoa: Eimeriidae) in cats. J. Protozool. 19:89–92.

Dubey, J.P. and J.K. Frenkel. 1972a. Cyst-induced toxoplasmosis in cats. J. Protozool. 19:155–77.

Dubey, J.P. and E.A. Hoover. 1977. Attempted transmission of *Toxoplasma gondii* infection from pregnant cats to their kittens. J. Am. Vet. Med. Assoc. 170:538–40.

Dubey, J.P. and J.L. Mahrt. 1979 (1978). *Isospora neorivolta* sp. n. from the domestic dog. J. Parasit. 64:1067–73.

Dubey, J.P. and H. Mehlhorn. 1978. Extraintestinal stages of *Isospora ohioensis* from dogs in mice. J. Parasit. 64:689–95.

Dubey, J.P., N.L. Miller, and J.K. Frenkel. 1970. The *Toxoplasma gondii* oocyst from cat feces. J. Exp. Med. 132:636–62.

Dubey, J.P., N.L. Miller, and J.K. Frenkel. 1970a. Characterization of the new fecal form of *Toxoplasma gondii.* J. Parasit. 56:447–56.

Dubey, J.P., N.L. Miller, and J.K. Frenkel. 1970b. The Toxoplasma gondii oocyst from cat feces. J. Exp. Med. 132:636–62.

Dubey, J.P. and B.P. Pande. 1963. Observations on the coccidian oocysts from Indian mongoose *(Herpestes mungo).* Ind. J. Microbiol. 3:49–54.

Dubey, J.P. and B.P. Pande. 1963a. Observations on the coccidian oocysts from Indian jungle cat *(Felis chaus)*. Ind. J. Microbiol. 3:103–8.

Dubey, J.P. and B.P. Pande. 1964. Letter to the editor. Ind. J. Microbiol. 4:29.

Dubey, J.P. and R.H. Streitel. 1976. *Isospora felis* and *I. rivolta* infections in cats induced by mice or oocysts. Brit. Vet. J. 132: 649–51.

Dubey, J.P. and R.H. Streitel. 1976a. Shedding of *Sarcocystis* in feces of dogs and cats fed muscles of naturally infected food animals in the midwestern United States. J. Parasit. 62:828–30.

Dubey, J.P. and R.H. Streitel. 1976b. Further studies on the transmission of *Hammondia hammondi* in cats. J. Parasit. 62:548–51.

Dubey, J.P., R.H. Streitel, P.C. Stromberg, and M.J. Toussant. 1977. *Sarcocystis fayeri* n. sp. from the horse. J. Parasit. 63:443–47.

Dubey, J.P., S.E. Weisbrode, and W.A. Rogers. 1978. Canine coccidiosis attributed to an *Isospora ohioensis*-like organism: A case report. J. Am. Vet. Med. Assoc. 173:185–91.

Dubey, J.P. and M.M. Wong. 1978. Experimental *Hammondia hammondi* infection in monkeys. J. Parasit. 64:551–52.

Dubremetz, J.-F., E. Porchet-Henneré, and M.-D. Parenty. 1975. Croissance de *Sarcocystis tenella* en culture cellulaire. C. R. Acad. Sci. 280(D):1793–95.

Dunlap, J.S. 1956. Experimental coccidial infection in the fox. West. Vet. 1956:64–67.

Duszynski, D.W. and C.A. Speer. 1976. Excystation of *Isospora arctopitheci* Rodhain, 1933 with notes on a similar process in *Isospora bigemina* (Stiles, 1891) Lühe, 1906. Z. Parasitenk. 48: 191–97.

Eichenwald, H.F. 1956. The laboratory diagnosis of toxoplasmosis. Ann. N. Y. Acad. Sci. 64:207–14.

Eisenstein, R. and J.R.M. Innes. 1956. Sarcosporidiosis in man and animals. Vet. Rev. Annot. 2:61–78.

Entzeroth, R., E. Scholtyseck, and E. Greuel. 1978. The roe deer intermediate host of different coccidia. Naturwissenschaften 65:395.

Erber, M. 1977. Moglichkeiten des Nachweises und der Differenzierung von zwei Sarcocystis-arten des Schweines. Berl. Münch. Tierärztl. Wschr. 90:480–82.

Erber, M. 1978. *Sarcocystis* spp. of wild boar and roe deer. Short. Com. Fourth Internat. Congr. Parasit. B:79.

Erber, M., J. Meyer, and J. Boch. 1978. Aborte beim Schwein durch Sarkosporidien *(Sarcocystis suicanis).* Berl Münch. Tierärztl. Wochenschr. 91:393–95.

Er-Hsiang, L. 1957. Finding of *Besnoitia besnoiti* in cattle in Peking. Acta Zool. Sinica 9:219.

Fameree, L. and C. Cotteleer. 1976. Toxoplasmose et hygiène. Prévalence de la coccidiose féline en Belgique. J. Protozool. 23:10A.

Farmer, J.N., I.V. Herbert, M. Partridge, and G.T. Edwards. 1978. The prevalence of *Sarcocystis* in dogs and red foxes. Vet. Rec. 102:78–80.

Fayer, R. 1972. Cultivation of feline *Isospora rivolta* in mammalian cells. J. Parasit. 58:1207–1208.

Fayer, R. 1974. Development of *Sarcocystis fusiformis* in the small intestine of the dog. J. Parasit. 60:660–65.

Fayer, R. 1974a. *Canis latrans* and *C. familiaris:* Host of *Sarcocystis fusiformis.* Proc. Third Internatl. Congr. Parasit. 1:114–15.

Fayer, R. 1977a. Production of *Sarcocystis cruzi* sporocysts by dogs fed experimentally infected and naturally infected beef. J. Parasit. 63:1072–75.

Fayer, R. 1977b. The first asexual generation in the life cycle of *Sarcocystis cruzi.* Proc. Helm. Soc. Wash. 44:206–9.

Fayer, R. and J.K. Frenkel. 1979. Comparative infectivity for calves of oocysts of feline coccidia: *Besnoitia, Hammondia, Cystoisospora, Sarcocystis,* and *Toxoplasma.* J. Parasit. 65:756–62.

Fayer, R. and A.J. Johnson. 1973. Development of *Sarcocystis fusiformis* in calves infected with sporocysts from dogs. J. Parasit. 59:1135–37.

Fayer, R. and A.J. Johnson. 1974. *Sarcocystis fusiformis:* Development of cysts in calves infected with sporocysts from dogs. Proc. Helminth. Soc. Wash. 41:105–8.

Fayer, R. and A.J. Johnson. 1975. *Sarcocystis fusiformis* infection in the coyote *(Canis latrans).* J. Infect. Dis. 131:189–92.

Fayer, R., A.J. Johnson, and P.K. Hildebrandt. 1976. Oral infection of mammals with *Sarcocystis fusiformis* bradyzoites from cattle and sporocysts from dogs and coyotes. J. Parasit. 62:10–14.

Fayer, R., A.J. Johnson, and M.N. Lunde. 1976. Abortion and other signs of disease in cows experimentally infected with *Sarcocystis* from dogs. J. Infect. Dis. 134:624–28.

Fayer, R. and D. Kradel. 1977. *Sarcocystis leporum* in cottontail rabbits and its transmission to carnivores. J. Wildl. Dis. 13:170–73.

Fayer, R. and R.G. Leek. 1973. Excystation of *Sarcocystis fusiformis* sporocysts from dogs. Proc. Helminth. Soc. Wash. 40: 294–96.

Fayer, R. and J.L. Mahrt. 1972. Development of *Isospora canis* (Protozoa: Sporozoa) in cell culture. Z. Parasitenk. 38:313–18.

Fayer, R., J.L. Mahrt, and A.J. Johnson. 1973. The life cycle and pathogenic effects of *Sarcocystis fusiformis* in experimentally infected calves (Film). J. Protozool. 20:509.

Fayer, R. and D.E. Thompson. 1974. *Isospora felis:* Development in cultured cells with some cytological observations. J. Parasit. 60:160–68.

Fedoseenko, V.M. and A.V. Levit. 1979. Elektronnomikroskopicheskoe izuchenne tsist *Sarcocystis muris* v skeletnoi muskulature belykh myshei. [Electron microscopic study of the cyst of *Sarcocystis muris* in skeletal muscles of white mice.] *In* Beyer, T.V. et al., eds. *Toksoplasmidy.* USSR Acad. Sci., Soc. Protozool. USSR, Ser. *Protozoology* No. 4. Izdat. Nauka, Leningrad, USSR. pp. 106–10.

Feldman, H.A. 1953. The clinical manifestations and laboratory diagnosis of toxoplasmosis. Am. J. Trop. Med. Hyg. 2:420–28.

Feldman, H.A. and L.T. Miller. 1956. Congenital human toxoplasmosis. Ann. N. Y. Acad. Sci. 64:180–84.

Fischer, G. 1979. Die Entwicklung von *Sarcocystis capracanis* n. spec. in der Ziege. D.M.V. thesis, Freien University Berlin. 45 p.

Flatt, R.E., L.R. Nelson, and M.M. Patton. 1971. *Besnoitia darlingi* in the opossum *(Didelphis marsupialis).* Lab. An. Sci. 21:106–9.

Ford, G.E. 1974. Prey-predator transmission in the epizootiology of ovine sarcosporidiosis. Austral. Vet. J. 50:38–39.

Ford, G.E. 1975. Transmission of sarcosporidiosis from dogs to sheep maintained specific pathogen free. Austral. Vet. J. 51:408.

Foreyt, W.J. and A.C. Todd. 1976. Prevalence of coccidia in domestic mink in Wisconsin. J. Parasit. 62:496.

Franti, C.E., G.E. Connolly, H.P. Riemann, D.E. Behymer, R. Ruppanner, C.M. Willadsen, and W. Longhurst. 1975. A survey for *Toxoplasma gondii* antibodies in deer and other wildlife on a sheep range. J. Am. Vet. Med. Assoc. 167:565–68.

Frelier, P., I.G. Mayhew, R. Fayer, and M.N. Lunde. 1977. Sarcocystosis: A clinical outbreak in dairy calves. Science 195:1341–42.

Frenkel, J.K. 1948. Dermal hypersensitivity to *Toxoplasma* antigens (toxoplasmins). Proc. Soc. Exper. Biol. Med. 68:634–39.

Frenkel, J.K. 1949. Pathogenesis, diagnosis and treatment of human toxoplasmosis. J. Am. Med. Assoc. 140:369–77.

Frenkel, J.K. 1970. Pursuing Toxoplasma. J. Infect. Dis. 122:553–59.

Frenkel, J.K. 1973. Toxoplasmosis: Parasite life cycle, pathology, and immunology. *In* Hammond, D.M. with P.L. Long, eds. *The Coccidia*. University Park Press, Baltimore, Md., pp. 343–410.

Frenkel, J.K. 1974. Breaking the transmission chain of *Toxoplasma:* A program for the prevention of human toxoplasmosis. Bull. N. Y. Acad. Med., 2nd Ser. 50:228–35.

Frenkel, J.K. 1974a. Advances in the biology of Sporozoa. Z. Parasitenk. 45:125–62.

Frenkel, J.K. 1975. Toxoplasmosis in cats and man. Fel. Pract. 1975:28–41.

Frenkel, J.K. 1977. *Besnoitia wallacei* of cats and rodents: With a reclassification of other cyst-forming isosporoid coccidia. J. Parasit. 63:611–28.

Frenkel, J.K. and J.P. Dubey. 1972. Rodents as vectors for feline coccidia, *Isospora felis* and *I. rivolta*. J. Infect. Dis. 125:69–72.

Frenkel, J.K. and J.P. Dubey. 1975. *Hammondia hammondi* gen. nov., sp. nov., from domestic cats, a new coccidian related to *Toxoplasma* and *Sarcocystis*. Z. Parasitenk. 46:3–12.

Frenkel, J.K. and J.P. Dubey. 1975a. *Hammondia hammondi:* A new coccidium of cats producing cysts in muscle of other mammals. Science 189:222–24.

Frenkel, J.K., J.P. Dubey, and N.L. Miller, 1970. *Toxoplasma gondii* in cats. Fecal stages identified as coccidian oocysts. Science 167:893–96.

Galli-Valerio, B. 1929. Notes de parasitologie. Zentralbl. Bakt. I. Abt. Orig. 112:54–59.

Galli-Valerio, B. 1931. Notes de parasitologie. Zentralbl. Bakt. I. Abt. Orig. 120:98–106.

Galli-Valerio, B. 1932. Notes de parasitologie et de technique parasitologique. Zentralbl. Bakt. I. Abt. Orig. 125:129–42.

Galli-Valerio, B. 1933. Notes de parasitologie et de technique parasitologique. Zentralbl. Bakt. I. Abt. Orig. 129:422–33.

Galli-Valerio, B. 1935. Parasitologische Untersuchungen und parasitologische Technik. Zentralbl. Bkt. I. Abt. Orig. 135:318–27.

Garnham, P.C.C. 1966. *Malaria Parasites and Other Haemosporidia.* Blackwell Scientific Publications, Oxford, 1132 p.

Gassner, F.X. 1940. Studies in canine coccidiosis. J.Am. Vet. Med. Assoc. 96:225–29.

Gestrich, R. 1974. Investigations on survival and resistance of *Sarcocystis fusiformis* cysts in beef. Proc. III Internatl. Congr. Parasit. 1:117.

Gestrich, R. and A.-O. Heydorn. 1974. Untersuchungen zur Uberlebensdauer von Sarkosporidienzysten im Fleisch von Schlachttieren. Berl. Münch. Tierärztl. Wochenschr. 87:475–76.

Gestrich, R., A.-O. Heydorn, and N. Baysu. 1975. Beitrage zum Lebenszyklus der Sarkosporidien. VI. Untersuchungen zur Artendifferenzierung bei Sarcocystis fusiformis und Sarcocystis tenella. Berl. Münch. Tierärztl. Wochenschr. 88:191–97, 201–4.

Gestrich, R., H. Mehlhorn, and A.-O. Heydorn. 1975. Licht- und elektronenmikroskopische Untersuchungen Cysten von Sarcoystis fusiformis in der Muskulatur von Kälbern nach experimenteller Infektion mit Oocysten und Sporocysten der grossen Form von Isospora bigemina der Katze. Zbl. Bakt. Hyg. I Abt. Orig. A 223:261–76.

Gestrich, R., M. Schmitt, and A.-O Heydorn. 1974. Pathogenität von Sarcocystis tenella-Sporozystern aus den Fäzes von Hunden für Lämmer. Berl. Munch. Tierärztl. Wochenschr. 87:362–63.

Ghaffar, F.A., M. Hilali, and E. Scholtyseck. 1978. Ultrasturctural study of Sarcocystis fusiformis (Railliet, 1897) infecting the Indian water buffalo (Bubalus bubalis) of Egypt. Tropen. Med. Parasit. 29:289–94.

Gill, H.A., A. Singh, D.V. Vadehra, and S.K. Sethi. 1978. Shedding of unsporulated *Isospora* oocysts in feces by dogs fed diaphragm muscles from water buffalo *(Bubalus bubalis)* naturally infected with *Sarcocystis.* J. Parasit. 64:549–51.

Göbel, E. 1976. Elektronenmikroskopische Untersuchungen zur Feinstrukter der Zystenstadien von Pferdesarkosporidien *(Sarcocystis equicanis).* Z. Parasitenk. 50:201.

Goldman, M., R.K. Carver, and A.J. Sulzer. 1957. Similar internal morphology of *Toxoplasma gondii* and *Besnoitia jellisoni* stained with silver protein. J. Parasit. 43:490–99.

Golemansky, V. 1975. *Eimeria li* nov. sp. et *Klossia* sp. (Protozoa:

Coccidia) trouves dans le gros intestin du renard commune (*Vulpes vulpes* L.) en Bulgarie. Zool. Anz. 194:133–39.

Golemanski, V. and N. Ridzhakov. 1975. V'rkho koktsidiite (Protozoa, Coccidia) na lisitsite v B'lgariya. Acta Zool. Bulgar. 3:3–18.

Golubkov, V.I. 1979. Zarazhenie sobak i koshek sarcotsistami ot kur i vtok. [Infection of the dog and cat with sarcocysts from the chicken and duck.] Veterinariya 1979(1):35–36.

Golubkov, V.I., O.V. Rybaltovskii, and Z.I. Kislyakova. 1974. Plotoyadnye—istochnik zarasheniya svinei sarkotsistami. Veterinariya 1974(II):85–86.

Goodrich, H.P. 1944. Coccidian oocysts. Parasitology 36:72–79.

Gousseff, W.F. 1933. Zur Frage der Coccidien der Füchse in Transkaukasien. Arch. Tierheilk. 66:424–428.

Gousseff, W.F. 1933a. Zur Frage der Katzenkokzidiose in Transkaukasien. Deut. Tierarztl. Wochenschr. 41:694–695.

Gräfner, G., H.-D. Graubmann, and W. Dobbriner. 1967. Leberkokzidiose beim Nerz (Lutreola vison Schreb.), hervorgerufen durch eine neue Kokzidienart, Eimeria hiepei n. sp. Monatsh. Vet.-Med. 22:696–700.

Grassé, P.P. 1953. Classe des sarcosporidies. Sarcosporidia Bütschli, 1882. *In* Grassé, P.P., ed. *Traité de Zoologie.* Masson, Paris. 1(2):907–17.

Grassi, G.B. 1879. Intorno a speciali corpuscoli (psorospermi) dell' uomo. Rendic. R. Ist. Lomb. Sci. Lett., 2nd Ser. 12:632–37.

Guterbock, W.M. and N.D. Levine. 1977. Coccidia and intestinal nematodes of east central Illinois cats. J. Am. Vet. Med. Assoc. 170:1411–1413.

Hair, J.D. and J.L. Mahrt. 1970. *Eimeria albertensis* n. sp. and *E. borealis* n. sp. (Sporozoa: Eimeriidae) in black bears *Ursus americanus* from Alberta. J. Protozool. 17:663–64.

Hartley, W.J. and W.F. Blakemore. 1974. An unidentified sporozoan encephalomyelitis in sheep. Vet. Path. 11:1–12.

Hartley, W.J. and J.C. Kater. 1963. The pathology of *Toxoplasma* infection in the pregnant ewe. Res. Vet. Sci. 4:326–32.

Hartley, W.J. and J.C. Kater. 1964. Perinatal disease conditions of sheep in New Zealand. N. Zeal. Vet. J. 12:49–57.

Hartley, W.J. and S.C. Marshall. 1957. Toxoplasmosis as a cause of ovine prenatal mortality. N. Zeal. Vet. J. 5:119–24.

Hasselmann, G. 1926. Alterações patologicas do myocardio na sarcosporideose. Bol. Inst. Brasil. Sci. 2:310–26.

Hendricks, L.D. 1977. Host range characteristics of the primate coccidian, *Isospora arctopitheci* Rodhain 1933 (Protozoa: Eimeriidae). J. Parasit. 63:32–35.

Hendricks, L.D., J.V. Ernst, C.H. Courtney, and C.A. Speer. 1979. *Hammondia pardalis* sp. n. (Protozoa: Eimeriidae) from the ocelot, *Felis pardalis*, and experimental infection of other felines. J. Protozool. 26:39–43.

Henry, A.C.L. 1913. [*Besnoitia besnoiti* (Marotel, 1913)]. Rec. Med. Vet. 90:328.

Henry, A. and C. Leblois. 1926. Essai de classification des coccidies de la famille des Diplosporidae Léger, 1911. Ann Parasit. 4:22–28.

Hepding, L. 1939. Ueber Toxoplasmen (*Toxoplasma gallinarum* n. sp.) in der Retina eines Huhnes und über deren Beziehung zur Hühnerlähmung. Z. Infektionskr. Haust. 55:109–16.

Hérin, V.V. 1952. Note sur l'existence de la globidiose au Ruanda-Urundi. Ann. Soc. Belge Med. Trop. 32:155–59.

Heydorn, A.-O. 1973. Zur Lebenszyklus der kleinen Form von *Isospora bigemina* des Hundes. I. Rind und Hund als mögliche Zwischenwirte. Berl. Münch. Tierärztl. Wochenschr. 86:323–29.

Heydorn, A.-O. and R. Gestrich. 1976. Beitrage zum Lebenszyklus der Sarkosporidien. VII. Entwicklungsstadien von Sarcocystis ovicanis im Schaf. Berl. Münch. Tierärztl. Wochenschr. 89:1–5.

Heydorn, A.-O., R. Gestrich, and K. Janitschke. 1976. Beitrage zum Lebenszyklus der Sarkosporidien. VIII. Sporozysten von Sarcocystis bovihominis in den Fäzes von Rhesusaffen (Macaca rhesus) und Pavianen (Papio cynocephalus). Berl. Münch. Tierärztl. Wochenschr. 89:116–20.

Heydorn, A.-O., R. Gestrich, H. Mehlhorn, and M. Rommel. 1975. Proposal for a new nomenclature of the *Sarcosporidia*. Z. Parasitenk. 48:73–82.

Heydorn, A.-O., H. Mehlhorn, and R. Gestrich. 1975. Lichtund elektronenmikroskopische Untersuchungen an Cysten von Sarcocystis fusiformis in der Muskulatur von Kälbern nach experimenteller Infektion mit Oocysten und Sporocysten von Isospora hominis Railliet et Lucet, 1891. 2. Die Feinstruktur der Metrocyten und Merozoiten. Zentralbl. Bakt. I. Orig. A 232:373–91.

Heydorn, A.-O. and M. Rommel. 1972. Beitrage zum Lebenzyklus der Sarkosporidien. II. Hund and Katze als Uberträger der Sarkosporidien des Rindes. Berl. Münch. Tierärztl. Wochenschr. 85: 121-23.

Heydorn, A.-O. and M. Rommel. 1972a. Beitrage zum Lebenszyklus der Sarkosporidien. IV. Entwicklungsstadien von S. fusiformis in der Dünndarmschleimhaut der Katze. Berl. Münch. Tierärztl. Wochenschr. 85:333-36.

Hilali, M. and E. Scholtyseck. 1978. The fine structure of the sexual stages of *Sarcocystis fusiformis* in domestic cats. Short Com. Fourth Internat. Congr. Parasit. B:75-76.

Hilgenfeld, M. and G. Punke. 1974. Zur Sarkosporidieninfektion des Zentralnervensystems des Schafes, ein Beitrag zur Differentialdiagnose von Protozoeninfektionen. Arch. Exp. Vet. Med. 28:621-24.

Hitchcock, D.J. 1955. The life cycle of *Isospora felis* in the kitten. J. Parasit. 41:383-98.

Hoare, C.A. 1927. On the coccidia of the ferret. Ann. Trop. Med. Parasit. 21:313-20.

Hofmeyr, C.F.B. 1945. Globidiosis in cattle. J. S. Afr. Vet. Med. Assoc. 16:102-9.

Honess, R.F. 1956. Parasites and parasitic diseases. Wyo. Game and Fish Com. Bull. No. 9 (Diseases of Wildlife in Wyoming):65-253.

Hudkins, G. and T.P. Kistner. 1977. *Sarcocystis hemionilatrantis* (sp. n.) life cycle in mule deer and coyotes. J. Wildl. Dis. 13:80-84.

Hudkins-Vivion, G., T.P. Kistner, and R. Fayer. 1976. Possible species differences between *Sarcocystis* from mule deer and cattle. J. Wildl. Dis. 12:86-87.

Hussein, M.F. and E.M. Haroun. 1975. Bovine cutaneous besnoitiosis in the Sudan: a case report. Brit. Vet. J. 131:85-88.

Hutchison, W.M., J.F. Dunachie, J.C. Siim, and K. Work. 1970. Coccidianlike nature of *Toxoplasma gondii*. Brit. Med. J. 1970: 1:142-44.

Hutchison, W.M., J.F. Dunachie, K. Work, and J.C. Siim. 1971. The life cycle of the coccidian parasite, *Toxoplasma gondii*, in the domestic cat. Trans. Roy. Soc. Trop. Med. Hyg. 65:380-99.

Inabnit, R., B. Chobotar, and J.V. Ernst. 1972. *Eimeria procyonis* sp. n., an *Isospora* species, and a redescription of *Eimeria nuttalli*

Yakimoff and Matikaschwili, 1932 (Protozoa: Eimeriidae) from the American raccoon (*Procyon lotor*). J. Protozool. 19:244–47.

Iseki, M., K. Tanabe, S. Uni, R. Sano, and S. Takada. 1974. A survey on Toxoplasma and other protozoal and helminthic parasites of adult stray cats in Osaka area. Jap. J. Parasit. 23:317–22.

Ito, S., K. Tsunoda, H. Nishikawa, and T. Matsui. 1974. Small type of *Isospora bigemina*: Isolation from naturally infected cats and relations with *Toxoplasma* oocyst. Nat. Inst. Anim. Health Quart. 14:137–44.

Ito, S., K. Tsunoda, and K. Shimura. 1978. Life cycle of the large type of *Isospora bigemina* of the cat. Nat. Inst. Anim. Health Quart. 18:69–82.

Iwanoff-Gobzem, P.S. 1934. Zum Vorkommen von Coccidien bei kleinen wilden Säugetieren. Deut. Tierärztl. Wochenschr. 42: 149–51.

Ivanova-Gobzem, P.S. 1935. K voprosy o koktsidiyakh domashnikh i dikikh zhivotnykh severnogo Kazakhstana. Vrediteli sel'skokh. Zhivot. Bor'ba Nimi. Po materalam parasitologicheskoi ekspeditsii v Severnom Kazakhstana 1932 r., ed. by E.N. Pavlovsky. Izdat. Akad. Nauk SSSR, pp. 243–63.

Jacob, E. 1949. Zur Übertragbarkeit der Coccidiose zwischen Hauskatze und Fuchs. Berl. Münch. Tierärztl. Wochenschr. 5:57.

Jacobs, L. 1957. The interrelation of toxoplasmosis in swine, cattle, dogs, and man. Publ. Health Repts. 72:872–82.

Jacobs, L. and M.N. Lunde. 1957. A hemagglutination test for toxoplasmosis. J. Parasit. 43:308–14.

Jacobs, L. 1970. Toxoplasmosis: Epidemiology and medical importance. J. Wildl. Dis. 6:305–12.

Janitschke, K., D. Protz, and H. Werner. 1976. Beitrag zum Entwicklungszyklus von Sarkosporidien der Grantgazelle (*Gazella granti*). Z. Parasitenk. 48:215–19.

Janitschke, K. and H. Werner. 1972. Untersuchungen ueber die Wirtsspezificitaet des geschlechtlichen Entwicklungszyklus von *Toxoplasma gondii*. Z. Parasitenk. 39:247–54.

Jensen, J.B. and S.A. Edgar. 1978. Fine structure of penetration of cultured cells by *Isospora canis* sporozoites. J. Protozool. 25: 169–73.

Jewell, M.I., J.K. Frenkel, K.M. Johnson, V. Reed, and A. Ruiz. 1972. Development of *Toxoplasma* oocysts in neotropical Felidae. Am. J. Trop. Med. Hyg. 21:512–17.

Jira, J. and V. Kozojed. 1970. *Toxoplasmose—1908–1967.* Fischer, Stuttgart, Germany. 2 vols. 396 p., 464 p.

Jones, S.R. 1973. Toxoplasmosis: A review. J. Am. Vet. Med. Assoc. 163:1038–42.

Kalyakin, V.N. and D.N. Zasukhin. 1975. Distribution of Sarcocystis (Protozoa: Sporozoa) in vertebrates. Fol. Parasit. 22:289–307.

Karstad, L. and D.O. Trainer. 1969. *Sarcocystis* in white-tailed deer. Bull. Wildl. Dis. Assoc. 5:25–26.

Khvan, M.V. 1969. Besnoitia besnoiti of cattle in Kazakhstan. Prog. Protozool. 3:232–33.

Kingscote, A.A. 1934. *Eimeria mustelae* n. sp., from *Mustela vison.* J. Parasit. 20:252–54.

Kingscote, A.A. 1935. A note on the coccidia of the mink. J. Parasit. 21:126.

Knowles, R. and B.M. Das Gupta. 1931. A note on two intestinal protozoa of the Indian mongoose. Ind. J. Med. Res. 19:175–76.

Kotlan, A. and L. Pospesch. 1933. Coccidial infections in the badger *Meles taxus* L. Parasitology 25:102–7.

Kozar, M. 1971. Diagnostyka mikroskopowa sarkosporidydiozy i czestosc wystepowania tej inwazji u niektorych zwierzat domowych w Polsce. Wiad. Parazyt. 17:29–39.

Kozar, Z. 1970. Toxoplasmosis and coccidiosis in mammalian hosts. *In* Jackson, G.J., R. Herman, and I. Singer, eds. *Immunity to Parasitic Animals.* Appleton-Century-Crofts, New York, 2:871–912.

Krause, C. and S. Goranoff. 1933. Ueber Sarkosporiosis bei Huhn und Wildente. Z. Infektskr. Haust. 43:361–78.

Kühn, J. 1865. Untersuchungen uber die Trichinenkrankheit der Schweine. Mitth. Landwirthsch. Inst. Univ. Halle 1865:1–84.

Labbé, A. 1899. Sporozoa. Das Tierreich, Berlin, 5 Lief., 200 p.

Lainson, R. 1957. The demonstration of *Toxoplasma* in animals, with particular reference to members of the Mustelidae. Trans. Roy. Soc. Trop. Med. Hyg. 51:111–17.

Lainson, R. 1968. Parasitological studies in British Honduras. III. Some coccidial parasites of mammals. Ann. Trop. Med. Parasit. 62:252–59.

Lainson, R. 1972. A note on Sporozoa of undetermined taxonomic position in an armadillo and a heifer calf. J. Protozool. 19:582–86.

Lankester, E.R. 1882. On *Drepanidium ranarum* the cell parasite of the frog's blood and spleen (Gaule's Wurmchen). Quart. J. Microsc. Sci. 22:53–65.

Lee, C.D. 1934. The pathology of coccidiosis in the dog. J. Am. Vet. Med. Assoc. 85:760–81.

Leek, R.G. and R. Fayer, 1979. Survival of sporocysts of *Sarcocystis* in various media. Proc. Helm. Soc. Wash. 46:151–54.

Leek, R.G., R. Fayer, and A.J. Johnson. 1977. Sheep experimentally infected with *Sarcocystis* from dogs. I. Disease in young lambs. J. Parasit. 63:642–50.

Léger, L. 1911. *Caryospora simplex,* coccidie monosporée et la classification des coccidies. Arch. Protistenk. 22:71–86.

Léger, L. and O. Duboscq. 1910. *Selenococcidium intermedium* Lég. et Dub. et la systématiques des sporozoaires. Arch. Zool. Exp. Gen. 5:187–238.

Lepp, D.L. and K.S. Todd, Jr. 1974. Life cycle of *Isospora canis* Nemeseri, 1959 in the dog. J. Protozool. 21:199–206.

Lepp, D.L. and K.S. Todd, Jr. 1976. Sporogony of the oocysts of *Isospora canis.* Trans. Am. Micro. Soc. 95:98–103.

Leuckart, R. 1879. *Die Parasiten des Menschen.* 2nd ed. G.F. Winter, Leipzig. 344 p.

Levaditi, S., V. Sanchis-Bayarri, P. Lépine, and R. Schoen. 1929. Etude sur l'encéphalomyélite provoquée par le *Toxoplasma cuniculi.* I, II. Ann. Inst. Pasteur 43:673–736, 1053–80.

Levine, N.D. 1948. *Eimeria* and *Isospora* of the mink *(Mustela vison).* J. Parasit. 34:486–92.

Levine, N.D. 1970. Taxonomy of the Sporozoa. J. Parasit. 56(II): 208–9.

Levine, N.D. 1973. *Protozoan Parasites of Domestic Animals and of Man.* 2nd ed. Burgess, Minneapolis, Minn. 413 p.

Levine, N.D. 1977a. Nomenclature of *Sarcocystis* in the ox and sheep and of fecal coccidia of the dog and cat. J. Parasit. 63: 36–51.

Levine, N.D. 1977b. Taxonomy of *Toxoplasma.* J. Protozool. 24: 36–41.

Levine, N.D. 1977c. Recent advances in classification of protozoa. Beltsville Symp. II, Biosystem. Agric. pp. 71–87.

Levine, N.D. 1978. *Textbook of Veterinary Parasitology.* Burgess, Minneapolis, Minn. 245 p.

Levine, N.D. and P.E. Cechner. 1973. The relationship of certain coccidia to *Sarcocystis* of swine. J. Protozool. 20:510–11.

Levine, N.D., P.E. Cechner, and R.C. Meyer. 1974. The relationship of certain coccidia to *Sarcocystis* of swine. Ind. Vet. J. 51: 57–59.

Levine, N.D. and V. Ivens. 1964. *Isospora spilogales* n. sp. and *I. sengeri* n. sp. (Protozoa: Eimeriidae) from the spotted skunk *Spilogales putorius ambarvalis.* J. Protozool. 11:505–9.

Levine, N.D. and V. Ivens. 1965a. *Isospora* species in the dog. J. Parasit. 51:859–64.

Levine, N.D. and V. Ivens. 1965b. *The Coccidian Parasites (Protozoa, Sporozoa) of Rodents.* Ill. Biol. Monogr. No. 33. University of Illinois Press, Urbana. 365 p.

Levine, N.D. and V. Ivens. 1970. *The Coccidian Parasites (Protozoa, Sporozoa) of Ruminants.* Ill. Biol. Monogr. No. 44. University of Illinois Press, Urbana. 278 p.

Levine, N.D., V. Ivens, and G.R. Healy. 1975. *Isospora herpestei* n. sp. (Protozoa, Apicomplexa) and other new species of *Isospora* from mongooses. Proc. Okla. Acad. Sci. 55:150–53.

Levine, N.D. and W. Tadros. 1980. Named species and hosts of *Sarcocystis* (Protozoa: Apicomplexa: Sarcocystidae). System. Parasit. (in press).

Lickfeld, K.G. 1959. Untersuchungen uber das Katzencoccid *Isospora felis* Wenyon, 1923. Arch. Protistenk. 103:427–56.

von Linstow, O.F.B. 1903. Parasiten, meistens Helminthen, aus Siam. Arch. Mikr. Anat. 62:108–21.

Litvenkova, A.E. 1969. Coccidia of wild mammals in Byelorussia. Prog. Protozool. 3:340–41.

Long, P.L. and C.A. Speer. 1977. Invasion of host cells by coccidia. *In* Taylor, A.E.R. and R. Muller, eds. *Parasite Invasion.* Blackwell Scientific Publications, Oxford, pp. 1–26.

Loveless, R.M. and F.L. Andersen. 1975. Experimental infection of coyotes with *Echinococcus granulosus, Isospora canis,* and *Isospora rivolta.* J. Parasit. 61:546–47.

Lühe, M. 1906. Die im Blute schmarotzenden Protozoen und ihre nächsten Verwandten. *In* Mense, C.A., ed. *Handbuch der Tropenkrankheiten, Anhang: Coccidia.* 3:69–268.

Luengo, M., N. Arata, and J. Luengo. 1974. Hallazgo de sarcosporidio en cerebelo de bovino. Bol. Chil. Parasit. 29:39–41.

Lunde, M.N. 1973. Laboratory methods in the diagnosis of toxoplasmosis. Health Lab. Sci. 10:319–28.

Lunde, M.N. and R. Fayer. 1977. Serologic tests for antibody to *Sarcocystis* in cattle. J. Parasit. 63:222–25.

Lunde, M.N. and L. Jacobs. 1967. Evaluation of a latex agglutination test for toxoplasmosis. J. Parasit. 53:933–36.

Machulskii, S.N. 1947. Trudy Buryat-Mong. Zoovet. Inst. 3:87–92. (*non vidi.* Cited by Kalyakin and Zasukhin, 1975.)

Machul'skii, S.N. and M.D. Miskaryan. 1958. Trudy Buryat-Mong. Zoovet. Inst. 13:297–99 (*non vidi.* Cited by Kalyakin and Zasukhin, 1975.)

Machul'skii, S.N. and P.S. Timofeev. 1940. Koktsidioz koshek v SSSR. Rev. Microbiol. Saratov 17:402–7.

Mackinnon, D.L. and M.J. Dibb. 1938. Report on intestinal protozoa of some mammals in the Zoological Gardens at Regent's Park. Proc. Zool. Soc. London 108 (B):323–45.

Mahrt, J.L. 1966. Life cycle of *Isospora rivolta* (Grassi, 1879) Wenyon, 1923 in the dog. Ph.D. thesis, University of Illinois, Urbana. 69 p.

Mahrt, J.L. 1967. Endogenous stages of the life cycle of *Isospora rivolta* in the dog. J. Protozool. 14:754–59.

Mahrt, J.L. 1968. Sporogony of *Isospora rivolta* oocysts from the dog. J. Protozool. 15:308–12.

Mahrt, J.L. 1973. *Sarcocystis* in dogs and its probable transmission from cattle. J. Parasit. 59:588–89.

Mahrt, J.L. 1973a. Cinemicrographic observations on the asexual development of *Isospora canis* in cultured cells. Prog. Protozool. 4:267.

Mahrt, J.L. and R. Fayer. 1975. Hematologic and serologic changes in calves experimentally infected with *Sarcocystis fusiformis.* J. Parasit. 61:967–69.

Mandal, L.N. and H.N. Ray. 1960. A new coccidium *Isospora leonina* n. sp. from a lion cub. Bull. Calcutta Sch. Trop. Med. 8:107–8.

Mandour, A.N. 1965. Life history of *Sarcocystis.* Prog. Protozool. 2:189–90.

Mantovani, A. 1965. Osservazioni sulla coccidiosi delle volpi in Abruzzo. Parassitologia 7:9–17.

Marchiondo, A.A., D.W. Duszynski, and G.O. Maupin. 1976. Prevalence of antibodies to *Toxoplasma gondii* in wild and domestic animals of New Mexico, Arizona, and Colorado. J. Wildl. Dis. 12:226–32.

Markus, M.B., R. Killick-Kendrick, and P.C.C. Garnham. 1974. The coccidial nature and life-cycle of *Sarcocystis.* J. Trop. Med. Hyg. 77:248–59.

Marotel, G. 1913. *[Sarcocystis besnoiti].* Rec. Med. Vet. 90:328.

Marotel, G. 1921. Sur une nouvelle coccidie du chat. Bull. Soc. Sci. Vet. Lyon 1921:86.

McCully, R.M., P.A. Basson, J.W. Van Niekerk, and R.D. Bigalke. 1966. Observations on *Besnoitia* cysts in the cardiovascular system of some wild antelopes and domestic cattle. Onderstepoort J. Vet. Res. 33:245–76.

McTaggart, H.S. 1960. Coccidia from mink in Britain. J. Parasit. 46:201–5.

Mehlhorn, H. 1974. Light and electron microscope studies on stages of Sarcocystis tenella in the intestine of cats. Proc. III Internat. Congr. Parasit. 1:105–6.

Mehlhorn, H., B. Becker, and A.-O. Heydorn. 1978. Light and electron microscopical studies on experiments producing gamogony of 5 *Sarcocystis* species in vivo and in tissue cultures. 2nd Germ. Jap. Coop. Symp. Prot. Dis. pp. 33–38.

Mehlhorn, H., W.J. Hartley, and A.-O. Heydorn. 1976. A comparative ultrastructural study of the cyst wall of 13 Sarcocystis species. Protistologica 12:451–67.

Mehlhorn, H., A.-O. Heydorn, and R. Gestrich. 1975. Lichtund elektronenmikroskopische Untersuchungen an Cysten von Sarcocystis fusiformis in der Muskulatur von Kälbern nach experimenteller Infektion mit Oocysten und Sporocysten der grossen Form von Isospora bigemina des Hundes. 1. Zur Entstehung der Cysten und der "Cystenwand." Zentralbl. Bakt. I. Orig. A 232: 392–409.

Mehlhorn, H. and M.B. Markus. 1976. Electron microscopy of stages of *Isospora felis* of the cat in the mesenteric lymph node of the mouse. Z. Parasitenk. 51:15–24.

Mehlhorn, H. and E. Scholtyseck. 1973. Elektronenmikroskopische Untersuchungen an Cystenstadien von *Sarcocystis tenella* aus der Oesophagus-Muskulatur des Schafes. Z. Parasitenk. 41:291–310.

Mehlhorn, H. and E. Scholtyseck. 1973a. Cytochemistry of the toxoplasmatea *Sarcocystis, Frenkelia,* and *Besnoitia* at the ultrastructural level. Prog. Protozool. 4:275.

Mehlhorn, H. and E. Scholtyseck. 1974. Licht und elektronenmikroskopische Untersuchungen an Entwicklungsstadien von *Sarcocystis tenella* aus der Darmwand der Hauskatze. I. Die Oocysten und Sporocysten. Z. Parasitenk. 43:251–70.

Mehlhorn, H. and E. Scholtyseck. 1974a. Die Parasit-Wirtsbeziehungen bei verscheidenen gattungen der Sporozoen (Eimeria, Toxoplasma, Sarcocystis, Frenkelia, Hepatozoon, Plasmodium und Babesia) unter Anwendung spezieller Verfahren. Micros. Acta 75:429–51.

Mehlhorn, H. and E. Scholtyseck. 1974b. Cytology and cytochemistry of *Sarcocystis tenella.* Proc. Third Internat. Congr. Parasit. 1:46–47.

Mehlhorn, H., J. Sénaud, B. Chobotar, and E. Scholtyseck. 1975. The origin of micronemes and rhoptries. Z. Parasitenk. 45:227–36.

Mehlhorn, H., J. Sénaud, A.-O. Heydorn, and R. Gestrich. 1975. Comparison des ultrastructures des kystes de *Sarcocystis fusiforms* Railliet, 1897 dans la musculature du boeuf, après, infection naturelle et après infection expérimentale par des sporocystes d'*Isospora hominis* et par des sporocystes des grandes forme d'*Isospora bigemina* du chien et du chat. Protistologica 11:445–55.

de Mello, I.F. 1915. Preliminary note on a new haemogregarine found in the pigeon's blood. Ind. J. Med. Res. 3:93–94.

de Mello, I.F. 1935. On a toxoplasmid of *Fulica atra* L. with special reference to a probable sexuality of agametes. Proc. Ind. Acad. Sci. (B) 1:705–79.

Mello, U. 1910. Un cas de toxoplasmose du chien observé à Turin. Bull. Soc. Path. Exot. 3:359–63.

Mesnil, F. 1916. [*Isospora bigemina* var. *canivelocis*] . Bull. Inst. Pasteur 14:130.

Mikeladze, L. 1978. Results of investigation of animal coccidia in Georgia. Short Com. Fourth Internat. Congr. Parasit. H:68.

Miller, N.L., J.K. Frenkel, and J.P. Dubey. 1972. Oral infections with *Toxoplasma* cysts and oocysts in felines, other mammals, and in birds. J. Parasit. 58:928–37.

Mimioglu, M., H. Güralp, and F. Sayin. 1960. Akara köpelklerinde görülen parazit turleri ve bunlarin yayilis nisbeti. Vet. Fak. Ankara Univ. 6:53–68. (Abstr. in Biol. Abstr. 35: No. 29656.)

Minchin, E.A. 1903. The Sporozoa. In Lankester, E.R., ed. A Treatise on Zoology. London. 1(2):150–360.

Mirza, M.Y. 1970. Incidence and distribution of coccidia (Sporozoa: Eimeriidae) in mammals from Baghdad area. M.S. Thesis, Baghdad, Iraq. 195 p.

Mirzayans, A., A.H. Eslami, M. Anwar, and Sanjar. 1972. Gastrointestinal parasites of dogs in Iran. Trop. Anim. Health Prod. 4: 58–60.

Morgan, B.B. and E.F. Waller. 1940. A survey of the parasites of the Iowa cottontail. (Sylvilagus floridanus mearnsi). J. Wildl. Man. 4:21–26.

Moulé, L.T. 1886. Psorospermose des bovidés. Bull. Soc. Centr. Med. Vet. 40 (n. S. 4):694–96.

Moulé, L. 1888. Des sarcosporidies et de leur fréquence, principalement chez les animaux de boucherie. Mem. Soc. Sci. Arts Vitry-le-Francois. 14:3–42.

de Moura Costa, M.D. 1956. Isosporose do cão—com a descrição de uma nova variedade (Isospora bigemina Stiles, 1891 bahiensis n. var.). Bol. Inst. Biol. Bahia 3:107–12.

Müller, B.E.G., H. Mehlhorn, and E. Scholtyseck. 1973. Identification of the cyst stages of four Toxoplasmatea based on ultrastructural characteristics: Toxoplasma gondii, Sarcocytis tenella, Besnoitia jellisoni, and Frenkelia sp. Prog. Protozool. 4:285.

Mukherjea, A.K. and S.M. Krassner, 1965. A new species of coccidia (Protozoa: Sporozoa) of the genus Isospora Schneider, 1881, from the jackal, Canis aureus Linnaeus. Proc. Zool. Soc. Calcutta 18:35–40.

Munday, B.L., I.K. Barker, and M.D. Rickard. 1975. The developmental cycle of a species of Sarcocystis occurring in dogs and sheep with observations on pathogenicity in the intermediate host. Z. Parasitenk. 46:111–23.

Munday, B.L., J.D. Humphrey, and V. Kila. 1977. Pathology produced by, prevalence of, and probable life cycle of species of Sarcocystis in the domestic fowl. Avian Dis. 21:697–703.

Munday, B.L., R.W. Mason, W.J. Hartley, P.J.A. Presidente, and D. Obendorf. 1978. Sarcocystis and related organisms in Australian wildlife. I. Survey findings in mammals. J. Wildl. Dis. 14: 417–33.

Munday, B.L. and M.D. Rickard, 1974. Is *Sarcocystis tenella* two species? Austral. Vet. J. 50:558–59.

Neitz, W.O. 1965. A check-list and host-list of the zoonoses occurring in mammals and birds in South and South West Africa. Onderstepoort J. Vet. Res. 32:189–374.

Neméséri, L. 1959. Adatok a kutya coccidiosisakoz. I. *Isospora canis*. (To the coccidiosis of dog. I. *Isospora canis*). (Hungarian text; Russian and English summaries). Magy. Allat. Lapja. 14: 91–92.

Neméséri, L. 1960. Beitrage zur Aetiologie der Coccidiose der Hund. I. *Isospora canis* n. sp. Acta Vet. Acad. Sci. Hung. 10:95–99.

Nery-Guimaraes, F. and H.A. Lage. 1973. Prevalencia e ciclo evolutivo de *Isospora felis* Wenyon, 1923 e *I. rivolta* (Grassi, 1879) Wenyon, 1923 em gatos. Mem. Inst. Oswaldo Cruz 71:43–66.

Nery-Guimaraes, F. and H.A. Lage. 1973a. Infecção *per os* de gatos com formas vegetativas de *Toxoplasma gondii* Nicolle & Manceaux, 1909 sem produção de oocistos. Mem. Inst. Oswaldo Cruz 71:67–86.

Neuman, M. 1962. An outbreak of besnoitiosis in cattle. Refuah Vet. 19:105–6.

Neuman, M. 1962a. The experimental infection of the gerbil *(Meriones tristrami shawii)* with *Besnoitia besnoiti*. Refuah Vet. 19: 184–88.

Neuman, M. 1972. Serological survey of *Besnoitia besnoiti* (Marotel 1912) infection in Israel by immunofluorescence. Zentralbl. Vet. Med. B. 19:391–96.

Neuman, M. 1974. Cultivation of *Besnoitia besnoiti* Marotel, 1912, in cell culture. Tropenmed. Parasit. 25:243–49.

Neuman, M. and T.A. Noble. 1963. Symptomatology and pathology of experimental besnoitiasis in the golden hamster. J. Protozool. 10(Suppl.):34.

Neveu-Lemaire, M. 1912. Parasitologie des animaux domestiques. Maladies parasitaires non bacteriennes. Paris.

Nicolle, C. and L. Manceaux. 1908. Sur une infection a corps de Leishman (ou organismes voisins) du gondi. C. R. Acad. Sci. 147:763–66.

Nicolle, C. and L. Manceaux. 1909. Sur un protozoaire nouveau du gondi. C. R. Acad. Sci. 148:369–72. '

Nieschulz, O. 1924. Over een geval van *Eimeria* infectie bij een kat (*Eimeria felina* n. sp.). Tijdskr. Diergeneesk. 51:129–31.

Nieschulz, O. 1925. Zur Verbreitung von Isospora-Infektionem bei Hunden und Katzen in den Niederland. Zentralbl. Bakt. I. Orig. 94:137–41.

Nieschulz, O. and A. Bos. 1933. Ueber die Coccidien der Silberfüchse. Deut. Tierärztl. Wochenschr. 41:819–20.

Nukerbaeva, K.K. and S.K. Svanbaev. 1973. Koktsidii pushnykh zverei v Kazakhstane. [Coccidia of fur-bearing mammals in Kazakhstan.] Vestnik Sel'skokh. Nauki Kazakh. 1973 (12):50–54.

Nukerbaeva, K.K. and S.K. Svanbaev. 1974. K voprosu o spetsifichnosti koktsidii plotoyadnykh. Izvest. Akad. Nauk Kazakh. SSR, Ser. Biol. Nauk 12(2):35–40.

Overdulve, J.P. 1970. The identity of *Toxoplasma* Nicolle & Manceaux, 1909 with *Isospora* Schneider, 1881. Proc. Konikl. Nederl. Akad. Wetensch., Ser. C. 73:129–51.

Pak, S.M., V.V. Perminova, and N.V. Eshtokina. 1979. *Sarcocystis citelli vulpes* sp. n. iz zheltykh suslikov (*Citellus fulvus*). [*Sarcocystis citellivulpes* n. sp. in the yellow suslik *Spermophilus fulvus*.] *In* Beyer, T.V. et al., eds. *Toksoplasmidy.* USSR Acad. Sci., Soc. Protozool. USSR, Ser. *Protozoology* No. 4. Izdat. Nauka, Leningrad, USSR. pp. 111–14.

Panasyuk, D.E., V.M. Mintyugov, M.V. Pyatov, A.A. Zyablov, and V.G. Golovin. 1971. K vidovoi spetsifichnosti sarkosporidii. [Species preference in Sarcosporidia.] Veterinariya 1971(4): 65–67. (Transl. No. 43, Univ. Ill. Col. Vet. Med.)

Pande, B.P., B.B. Bhatia, and P.P.S. Chauhan. 1972. A new genus and species of cryptosporidid coccidia from India. Acta Vet. Acad. Sci. Hung. 22:231–34.

Pande, B.P., B.B. Bhatia, P.P.S. Chauhan, and R.K. Garg. 1970. Species composition of coccidia of some of the mammals and birds at the Zoological Gardens, Lucknow (Uttar Pradesh). Ind. J. An. Sci. 40:154–66.

Patnaik, M.M. and L.N. Acharjyo. 1970. *Eimeria nycticebi* n. sp. and *E. coucangi* n. sp. from Indian slow loris (*Nycticebus cou-*

cang) and notes on *Isospora leonina* from an African lion *(Panthera leo leo).* Orissa Vet. J. 5:13-14.

Patnaik, M.M. and S.K. Ray. 1965. Coccidia of Indian mongoose *(Herpestes edwardsii).* Ind. J. An. Health 4:33-36.

Paitnaik, M.M. and S.K. Ray. 1966. Letter to the editor. Ind. J. An. Health 5:203.

Pellérdy, L. 1955. Beiträge zur Kenntnis der Coccidien des Dachses *(Meles taxus).* Acta Vet. Acad. Sci. Hung. 5:421-24.

Pellérdy, L. 1963. *Catalogue of Eimeriidae (Protozoa; Sporozoa).* Akad. Kiado, Budapest, Hungary. 160 p.

Pellérdy, L.P. 1965. *Coccidia and Coccidiosis.* Akad. Kiado, Budapest. 657 p.

Pellérdy, L.P. 1969. *Catalogue of Eimeriidae (Protozoa, Sporozoa),* Suppl. I. Akad. Kiado, Budapest. 80 p.

Pellérdy, L.P. 1974. *Coccidia and Coccidiosis.* 2nd ed. Akad. Kiado, Budapest, and Paul Parey, West Berlin. 959 p.

Pelster, B. and G. Piekarski. 1971. Elektronenmikroskopische Analyse der Mikrogametenentwicklung bei *Toxoplasma gondii.* Z. Parasitenk. 37:267-77.

Peteshev, V.M., I.G. Galouzo and A.P. Polomoshnov. 1974. Koshki —definitivnye khozyaeva besnoitii *(Besnoitia besnoiti).* Izvest. Akad. Nauk Kazakh. SSR, Ser. Biol. 1:33-38. (Transl. No. 50, Univ. Ill. Col. Vet. Med.).

Peteshev, V.M., A.P. Polomoshnov, and N.V. Eshtokina. 1975. [*Citellus fulvus* as an experimental model for the study of *Besnoitia besnoiti.*] Voprosy Prirodnoi Ochagovosti Boleznei, Alma-Ata, USSR. 7:84-89. (Prot. Abstr. 1:213) *(Non vidi.)*

Pfefferkorn, E.R., L.C. Pfefferkorn, and E.D. Colby. 1997. Development of gametes and oocysts in cats fed cysts derived from cloned trophozoites of *Toxoplasma gondii.* J. Parasit. 63:158-59.

Piekarski, G., B. Pelster, and H.M. Witte. 1971. Endopolygenie bei *Toxoplasma gondii.* Z. Parasitenk. 36:122-30.

Piekarski, G. and H.M. Witte. 1971. Experimentelle und histologische Studien zur *Toxoplasma*-Infektion der Hauskatze. Z. Parasitenk. 36:95-121.

Poche, F. 1913. Das System der Protozoa. Arch. Protistenk. 30:125-321.

Pols, J.W. 1954. The artificial transmission of *Globidium besnoiti*

Marotel, 1912, to cattle and rabbits. J. S. Afr. Vet. Med. Assoc. 25:37–44.

Pols, J.W. 1960. Studies on bovine besnoitiosis with special reference to the aetiology. Onderstepoort J. Vet. Res. 28:266–356.

Porchet-Henneré, E. 1975. Quelques précisions sur l'ultrastructure de *Sarcocystis tenella*. I. L'endozoite (après coloration négative). J. Protozool. 22:214–20.

Porchet-Henneré, E. and G. Ponchel. 1974. Quelques précision sur l'ultra-structure de *Sarcocystis tenella*: L'architecture du kyste et l'aspect des endozoites en microscopie électronique 'a balayage. C.R. Acad. Sci. 279(D):1179–81.

Powell, E.C. and M.J. Last. 1977. Infections of both *Sarcocystis muris* and *Isospora felis* in mice used by Powell and McCarley (1975) in studying the life cycle of *S. muris*. Prog. Abstr. Amer. Soc. Parasit. 52:49–50.

Powell, E.C. and J.B. McCarley. 1975. A murine *Sarcocystis* that causes an *Isospora*-like infection in cats. J. Parasit. 61:928–31.

Prasad, H. 1961. A new species of *Isospora* from the fennec fox *Fennecus zerda* Zimmermann. Z. Parasitenk. 21:130–35.

Prasad, H. 1961a. The coccidia of the zorille *Ictonyx (Zorilla) capensis* Kaup. J. Protozool. 8:55–58.

Prestwood, A.K., S.R. Pursglove, and F.A. Hayes. 1976. Parasitism among white-tailed deer and domestic sheep on a common range. J. Wildl. Dis. 12:380–85.

von Prowazek, S. 1910. Parasitische Protozoen aus Japan, gesammelt von Herrn Dr. Mine in Fakuoka. Arch. Schiffshyg. 14:296–302.

Railliet, A. 1886. *[Miescheria tenella]*. Bull. Mem. Soc. Centr. Med. Vet. 40:130.

Railliet, A. 1886a. Psorospermies géantes dans l'oesophage et les muscles du mouton. Bull. Mem. Soc. Centr. Med. Vet. 40:130–34.

Railliet, A. 1897. La douve pancréatique. Bull. Soc. Centr. Med. Vet. 51:371–77.

Railliet, A. and A. Lucet. 1891. Note sur quelques espèces de coccidies encore peu étudiés. Bull. Soc. Zool. France 16:246–50.

Rajasekariah, G.R., K.S. Hegde, R.N.S. Gowda, S.A. Rahman, and H. Subbarao. 1971. A study of some parasites from panther cub

(*Felis pardus* Linn.), with description of *Eimeria anakalensis* n. sp. Mysore J. Agric. Sci. 5:404–9.

Rao, S.R. and M.Y. Bhatavdekar. 1957. *Eimeria rayii* sp. nov., a new coccidium from dog belonging to the genus *Eimeria*. Bombay Vet. Col. Mag. 6(Mar.):7–8.

Rastegaieff, E.F. 1929. *Eimeria mesnili* n. sp. chez *Canis lagopus*. Bull. Soc. Path. Exot. 22:640.

Rastegaieff, E.F. 1929a. *Eimeria felina* Niesch. chez la lionne. Bull. Soc. Path. Exot. 22:641.

Rastegaieff, E.F. 1929b. Coccidie chez le tigre. Bull. Soc. Path. Exot. 22:640.

Rastegaieff, E.F. 1930. Zur Frage uber Coccidien wilder Tiere. Arch Protistenk. 71:377–404.

Ratz, I. 1908. Ueber die in Muskeln parasitirenden Sarcosporidien und deren in Ungarns fauna vorkommende Arten. Allat. Közlem. 7:177–80.

Reichenow, E. 1953. *Lehrbuch der Protozoenkunde.* Gustav Fischer, Jena. 1213 p.

Remington, J.S. 1969. A common mechanism of immunity for intracellular infections. Prog. Protozool. 3:241–42.

Remington, J.S., P. Earle and T. Yagura. 1970. *Toxoplasma* in nucleus. J. Parasit. 56:390–91.

Remington, J.S., L. Jacobs, and H.E. Kaufman. 1960. Toxoplasmosis in the adult. New Engl. J. Med. 262:180–86, 237–41.

Roberts, W.L., J.L. Mahrt, and D.M. Hammond. 1972. The fine structure of the sporozoites of *Isospora canis*. Z. Parasitenk. 40: 183–94.

da Rocha, E.M. and C.W.G. Lopes. 1971. Comportemento da Isospora canis, Isospora felis e Isospora rivolta em infecções experimentais en cães e gatos. Arq. Univ. Fed. Rur. Rio de Janeiro 1: 65–70.

Rommel, M. 1975. Neue Erkenntnisse zur Biologie der Kokzidien, Toxoplasmen, Sarkosporidien und Besnoitien. Berl. Münch. Tierärztl. Wochenschr. 88:112–17.

Rommel, M. and O. Geisel. 1975. Untersuchungen über die Verbreitung und den Lebenszyklus einer Sarkosporidienart des Pferdes (Sarcosystis equicanis n. spec.). Berl. Münch. Tierärztl. Wochenschr. 88:468–71.

Rommel, M., A.-O. Heydorn, B. Fischle, and R. Gestrich. 1974. Beiträge zum Lebenszyklus der Sarkosporidien. V. Weitere Endwirte der Sarkosporidien von Rind, Schaf und Schwein und die Bedeutung des Zwischenwirtes fur die Verbreitung dieser Parasitose. Berl. Münch. Tierärztl. Wochenschr. 85:392–96.

Rommel, M., A.-O. Heydorn, and F. Gruber. 1972. Beiträge zum Lebenszyklus der Sarkosporidien. I. Die Sporozyte von S. tenella in den Fäzes der Katze. Berl. Münch. Tierärztl. Wochenschr. 85: 101–5.

Rommel, M. and F. von Seyerl. 1976. Der erstmalige Nachweis von Hammondia hammondi (Frenkel und Dubey 1975) im Kot einer Katze in Deutschland. Berl. Münch. Tierärztl. Wochenschr. 89:398–99.

Ruiz, A. and J.K. Frenkel. 1976. Recognition of cyclic transmission of Sarcocystis muris by cats. J. Infect. Dis. 133:409–18.

Rybaltovskii, O.V., A.V. Dudkina, and A.P. Rubina. 1973. K voprosy o tsikle razvitiya sarcotsist. [The life cycle of Sarcocystis.] Inst. Myasnoi Promyshlennosti, Moskva. 1973(11):71. (Transl. No. 54, Univ. Ill. Col. Vet. Med.).

Ryšavý, B. 1954. Prispevek k poznani koktsidii nasich i dovezenych obratlovcu. Csl. Parasit. 1:131–75.

Sabin, A.B. 1949. Complement fixation tests in toxoplasmosis and persistence of the antibody in human beings. Pediatrics 4:443–53.

Sabin, A.B., H. Eichenwald, H.A. Feldman, and L. Jacobs. 1952. Present status of clinical manifestations of toxoplasmosis in man. Indications and provisions for routine serologic diagnosis. J. Am. Med. Assoc. 150:1063–69.

Sabin, A.B. and H.A. Feldman. 1948. Dyes as microchemical indicators of a new immune phenomenon affecting a protozoan parasite (Toxoplasma). Science 108:660–63.

Sahasrabudhe, V.K. and H.L. Shah. 1966. The occurrence of Sarcocystis sp. in the dog. J. Protozool. 13:531.

Sangiorgi, G. 1913. Un nuovo protozoo parassita del Mus musculus. Pathologica 5:323–25.

Sangiorgi, G. 1915 (1914). Toxoplasma ratti n. sp. Gior. R. Accad. Med. Torino 77:383–85.

Schmitz, J.A. and W.W. Wolf. 1977. Spontaneous fatal sarcocystosis in a calf. Vet. Path. 14:527–31.

Schneider, A. 1875. Contributions a l'histoire des grégarines d'invertebres de Paris et de Roscoff. Arch. Zool. Exper. Gen. 4:493–604.

Schneider, A. 1881. Sur les psorospermies oviformes ou coccidies. Espèces nouvelles ou peu connues. Arch. Zool. Exper. Gen. 9: 387–403.

Schneider, C.R. 1965. *Besnoitia panamensis* sp. n. (Protozoa: Toxoplasmatidae) from Panamanian lizards. J. Parasit. 51:340–44.

Scholtyseck, E. 1973. Ultrastructure. *In* Hammond, D.M. with P.L. Long, eds. *The Coccidia*. University Park Press, Baltimore, Md., pp. 81–144.

Scholtyseck, E. and M. Hilali. 1978. Ultrastructural study of the sexual stages of *Sarcocystis fusiformis* (Railliet, 1897) in domestic cats. Z. Parasitenk. 56:205–9.

Scholtyseck, E., H. Mehlhorn, and B.E.G. Müller. 1973. Identifikation von Merozoiten der vier cystenbildenden Coccidien *(Sarcocystis, Toxoplasma, Besnoitia, Frenkelia)* auf Grund feinstruktureller Kriterien. Z. Parasitenk. 42:185–206.

Schrecke, W. and U. Dürr. 1970. Excystations und Infektionsversuche mit Kokzidienoocysten bei neugeborenen Tieren. Zentralbl. Bakt. I. Orig. 125:252–58.

Scott, J.W. 1943. Life history of Sarcosporidia, with particular reference to *Sarcocystis tenella*. Wyo. Agr. Expt. Sta. Bull. No. 259. 63 p.

Šebek, Z. 1975. Parasitische Gewebeprotozoen der wildlebenden Kleinsäuger in der Tschechoslowakei. Fol. Parasit. 22:111–24.

Sénaud, J. 1963. Les modalities de la multiplication des elements cellulaires dans les kystes de la sarcosporidie du mouton *(Sarcocystis tenella* Railliet, 1886). C. R. Acad. Sci. 256:1009–11.

Sénaud, J. 1967. Contribution a l'étude des sarcosporidies et des toxoplasmes *(Toxoplasmea)*. Protistologica 3:167–232.

Sénaud, J. and H. Mehlhorn. 1975. Etude ultrastructurale des coccidies formant des kystes: *Toxoplasma gondii, Sarcocystis tenella, Besnoitia jellisoni* et *Frenkelia* sp. *(Sporozoa)*. II. Mise en evidence de l'ADN et de l'ARN au niveau des ultrastructures. Ann. Sta. Biol. Besse-en-Chandesse 1975(9):111–56.

Seneviratna, P., A.G. Edward and D.L. DeGiusti. 1975. Frequency of *Sarcocystis* spp. in Detroit Metropolitan Area, Michigan. Am. J. Vet. Res. 36:337–339.

Shah, H.L. 1969. The coccidia (Protozoa: Eimeriidae) of the cat. Ph.D. Thesis, University of Illinois, Urbana. 148 p.

Shah, H.L. 1970. *Isospora* species of the cat and attempted transmission of *I. felis* Wenyon, 1923 from the cat to the dog. J. Protozool. 17:603–9.

Shah, H.L. 1970a. Sporogony of the oocysts of *Isospora felis* Wenyon, 1923 from the cat. J. Protozool. 17:609–14.

Shah, H.L. 1971. The life cycle of *Isospora felis* Wenyon, 1923, a coccidium of the cat. J. Protozool. 18:3–17.

Sheffield, H.G. and R. Fayer. 1978. Electron microscopy of in situ development of *Sarcocystis cruzi* oocysts. Short Com. 4th Internat. Congr. Parasit. B:77–78.

Sheffield, H.G., J.K. Frenkel, and A. Ruiz. 1977. Ultrastructure of the cyst of *Sarcocystis* muris. J. Parasit. 63:629–41.

Sheffield, H.G. and M. Melton. 1970. Toxoplasma gondii: The oocyst, sporozoite, and infection of cultured cells. Science 167: 892–93.

Sheffield, H.G. and M.L. Melton. 1974. Immunity to *Toxoplasma gondii* in cats. Proc. Third Internat. Congr. Parasit. 1:106–7.

Sheffield, H.G., M.L. Melton, and F.A. Neva. 1976. Development of *Hammondia hammondi* in cell cultures. Proc. Helm. Soc. Wash. 43:217–25.

Shelton, G.C., L.D. Kintner, and D.O. MacKintosh. 1968. A coccidia-like organism associated with subcutaneous granulomata in a dog. J. Am. Vet. Med. Assoc. 152:263–67.

Siedamgrotzky, O.A. 1872. Psorospermienschläuche in den Muskeln der Pferde. Wochenschr. Tierl. Viehsucht. 16:97–101.

Siim, J.C. 1956. Toxoplasmosis acquisita lymphonodosa: Clinical and pathological aspects. Ann. N. Y. Acad. Sci. 64:185–206.

Siim, J.C. 1974. *Toxoplasma gondii.* Actual. Protozool. 1:203–12.

Simitch, T., Z. Petrovitch, and A. Bordjochki. 1956. *Citellus citellus* animal de choix pur l'étude biologique et l'isolement de *Toxoplasma gondii.* Arch. Inst. Pasteur Algérie 34:93–99.

Simpson, C.F. 1966. Electron microscopy of *Sarcocystis fusiformis.* J. Parasit. 52:607–13.

Skibsted, S. 1945. Om forekomst af Sarkosporidier. Maanedsskr. Dyrlaeger 57:27–34.

Smith, D.D. and J.K. Frenkel. 1977. *Besnoitia darlingi* (Protozoa, Toxoplasmatidae): Cyclic transmission by cats. J. Parasit. 63: 1066–71.

Sominskii, Z.F., D.I. Panasyuk, and R.P. Vilkova. 1971. Patologo-morologicheskie izmeneniya pri eksperimental' nom sarkotsis-toze kur. [Pathological changes in experimental sarcocystosis of chickens.] Veterinariya 1971(6):68–69. (Transl. No. 44, Univ. Ill. Col. Vet. Med.).

Speer, C.A., D.M. Hammond, J.L. Mahrt, and W.L. Roberts. 1973. Structure of the oocyst and sporocyst walls and excystation of sporozoites of *Isospora canis*. J. Parasit. 59:35–40.

Splendore, A. 1908. Un nuovo protozoa parassita de' conigli incontrato nelle lesioni anatomiche d'una malattia ehe ricorda in molti punti il kala azar dell' uomo. Rev. Soc. Sci. S. Paulo 3:109–12.

Sprehn, C. and J. Cramer. 1931. Das Darmcoccid *Lucetina canivel-ocis* (Weidmann, 1915) in Silberfüchsen. Berl. Tierärztl. Wochenschr. 47:261–63.

Stiles, C.W. 1891. Note preliminaire sur quelques parasites. Bull. Soc. Zool. France 16:163.

Stiles, C.W. 1893. Notes on parasites—18: on the presence of sarcosporidia in birds. USDA BAI Bull. No. 3:79–85.

Streitel, R.H. and J.P. Dubey. 1976. Prevalence of *Sarcocystis* infection and other intestinal parasitisms in dogs from a humane shelter in Ohio. J. Am. Vet. Med. Assoc. 168:423–24.

Suteu, E. and S. Coman. 1973. Nouvelles observations sur le cycle biologique de *Sarcocystis fusiformis*. Bull. Soc. Sci. Vet. Med. Comp. 75:363–67.

Svanbaev, S.K. 1956. Materaly k faune koktsidii dikikh mlekopit-ayushchikh zapadnogo Kazakhstana. Trudy Inst. Zool. Akad. Nauk Kazakh. SSR 5:180–91.

Svanbaev, S.K. 1960. Koktsidii cerebristo-chernykh lisits Alma-Atinskoi oblasti. Trudy Inst. Zool. Akad. Nauk Kazakh. SSR 14:34–36.

Swellengrebel, N.H. 1914. Zur Kenntnis der Entwichlungsgeschichte von *Isospora bigemina* (Stiles). Arch. Protistenk. 32:379–92.

Tadros, W. and J.J. Laarman. 1976. *Sarcocystis* and related coccidian parasites: A brief general review, together with a discussion on some biological aspects of their life cycles and a new proposal for their classification. Acta Leidensia 44:1–107.

Tadros, W. and J.J. Laarman. 1977. The cat *Felis catus* as the final host of *Sarcocystis cuniculi* Brumpt, 1913 of the rabbit *Oryctolagus cuniculus*. Proc. Konink. Ned. Akad. Wetensch., Ser. C. 80:351–52.

Tadros, W. and J.J. Laarman. 1978. A comparative study of the light and electron microscopic structure of the walls of the muscle cysts of several species of sarcocystid eimeriid coccidia. Proc. Konink. Ned. Akad. Wetensch., Ser. C. 81:469–91.

Tadros, W. and J.J. Laarman. 1978a. Note on the specific designation of *Sarcocystis putorii* (Railliet and Lucet, 1891) comb. nov. of the common European vole, *Microtus arvalis.* Proc. Konink. Ned. Akad. Wetensch., Ser. C. 81:466–68.

Tadros, W. and J.J. Laarman. 1979. Muscular sarcosporidiosis in the common European weasel, *Mustela nivalis.* Z. Parasitenk. 58:195–200.

Thornton, J.E., R.R. Bell, and M.J. Reardon. 1974. Internal parasites of coyotes in southern Texas. J. Wildl. Dis. 10:232–36.

Tomimura, T. 1957. Experimental studies on coccidiosis in dogs and cats (1) The morphology of oocysts and sporogony of *Isospora felis* and its artificial infection in cats. Jap. J. Parasit. 6:12–24.

Torres, P., A. Hott, and H. Boehmwald. 1972. Protozoos, helmintos y artropodes en gatos de la ciudad de Valdivia y su importancia para el hombre. Arch. Med. Vet., Valdivia, Chile 4:2–11.

Trayser, C.V. 1973. Life cycle of *Isospora* species. M.S. Thesis, University of Illinois, Urbana, 46 p.

Trayser, C.V. and K.S. Todd, Jr. 1978. Life cycle of *Isospora burrowsi* n. sp. from the dog *(Canis familiaris).* Am. J. Vet. Res. 39:95–98.

Triffitt, M.J. 1927. Observations on the oocysts of coccidia found in the faeces of carnivores. Protozoology 3:59–64.

Vershinin, I.I. 1975. Sarkotsisty krupnogo rogatogo skota. Dokl. Vses. Akad. Sel'skokh. Nauk, Sverdlovsk 1975(1):130–32.

Vivier, E. 1970. Observations nouvelles sur la reproduction asexuée de *Toxoplasma gondii* et considérations sur la notion d'endogenese. C. R. Acad. Sci. 271(D):2124–26.

Vivier, E. and A. Petitprez. 1972. Données ultrastructurales complémentaires, morphologiques et cytochimiques, sur *Toxoplasma gondii.* Protistologica 8:199–221.

Vogelsang, E.G. 1938. Contribucion al estudio de la parasitologia animal en Venezuela. VIII. *Sarcocystis iturbei* sp. n. del bovino *(Bos taurus* L.). Bol. Soc. Venezolana Cien. Nat. 4:279–80.

Vsevolodov, B.P. 1961. O beznoitiotioze krupnogo rogatogo skota v Kazakhstane. (On besnoitiosis of cattle in Kazakhstan). *In*

Galuzo, E.G. et al., eds. *Prirodnaya Ochagovost' Bolezney i Voprosy*. (Translated by F.K. Plous, Jr., edited by N.D. Levine. 1968. *Natural Nidality of Diseases and Questions of Parasitology*. University of Illinois Press, Urbana, pp. 136–42).

Walker, E.P., F. Warnick, S.E. Hamlet, K.I. Lange, M.A. Davis, H.E. Uible, and P.F. Wright, revised by J.L. Paradiso. 1975. *Mammals of the World*. 3rd ed. Johns Hopkins University Press, Baltimore. 2 vols. 1549 p.

Wallace, G.D. 1971. Experimental transmission of *Toxoplasma gondii* by filth-flies. Am. J. Trop. Med. Hyg. 20:411–13.

Wallace, G.D. 1973. *Sarcocystis* in mice inoculated with *Toxoplasma*-like oocysts from cat feces. Science 180:1375–77.

Wallace, G.D. 1975. Observations on a feline coccidium with some characteristics of *Toxoplasma* and *Sarcocystis*. Z. Parasitenk. 46:167–78.

Wallace, G.D. and J.K. Frenkel. 1975. Besnoitia species (Protozoa, Sporozoa, Toxoplasmatidae): Recognition of cyclic transmission by cats. Science 188:369–71.

Walls, K.W. and I.G. Kagan. 1967. Studies on the prevalence of antibodies to *Toxoplasma gondii*. 2. Brazil. Am. J. Epidem. 86: 305–13.

Walls, K.W., I.G. Kagan, and A. Turner. 1967. Studies on the prevalence of antibodies to *Toxoplasma gondii*. 1. U.S. military recruits. Am. J. Epidem. 85:87–92.

Walton, A.C. 1959. Some parasites and their chromosomes. J. Parasit. 45:1–20.

Warren, J. and A.B. Sabin. 1943. The complement fixation reaction in toxoplasmosis. Proc. Soc. Exper. Biol. Med. 51:11–14.

Wasielewski, T. 1904. *Studien und Mikrophotogramme zur Kenntnis der pathogenen Protozoen*. Leipzig.

Watkins, C.V. and L.A. Harvey. 1942. On the parasites of silver foxes on some farms in the southwest. Parasitology 34:155–79.

Weidman, F.D. 1915. *Coccidium bigeminum* Stiles in swift-foxes (habitat western U.S.). J. Comp. Path. 28:320–23.

Weiland, G. and E. Kaggwa. 1976. Fluoreszens und enzymserologische Untersuchungen an experimentell mit *Besnoitia besnoiti* infizierten Kaninchen. Z. Parasitenk. 76:177.

Weiland, G. and D. Kühn. 1970. Experimentelle Toxoplasma-Infektionen bei der Katze. II. Entwicklungsstadien des Parasiten im Darm. Berl. Münch. Tierärztl. Wochenschr. 83:128–32.

Wenyon, C.M. 1923. Coccidiosis of cats and dogs and the status of *Isospora* of man. Ann. Trop. Med. Parasit. 17:231-88.

Wenyon, C.M. 1926. *Protozoology.* 2 vols. Wood, New York. 1579 p.

Wenyon, C.M. and L. Sheather. 1925. *Isospora* infections of dogs. Trans. Roy. Soc. Trop. Med. Hyg. 19:10.

Wetzel, R. 1938. Ein neues coccid (*Cryptosporidium vulpis* sp. nov.) aus dem Rotfuchs. Arch. Wiss. Prakt. Tierheilk. 74:39-40.

Willey, A., A.J. Chalmers, and W.M. Philip. 1904. Report on parasites in the carcasses of buffaloes at the Colombo slaughter house. Spolia Zeylan. 2:65-72.

Witte, H.M. and G. Piekarski. 1970. Die Oocysten—Ausscheidung bei experimentell infizierten Katzen in Abhangigkeit von Toxoplasma-Stama. Z. Parasitenk. 33:358-60.

Wolf, A., D. Cowen, and B.H. Paige. 1939. Human toxoplasmosis. Occurrence in infants as an encephalomyelitis. Verification by transmission to animals. Science 89:226-27.

Work, K. and W.M. Hutchison. 1969. A new cystic form of *Toxoplasma gondii.* Acta Path. Microbiol. Scand. 75:191-92.

Work, K. and W.M. Hutchison. 1969a. The new cyst of *Toxoplasma gondii.* Acta Path. Microbiol. Scand. 77:414-24.

Yakimoff, W.L. 1933. Zur Frage der Eimeriose der Katzen. Arch. Protistenk. 80:172-76.

Yakimoff, W.L. and W.F. Gousseff. 1934. Coccidia of martens and sables. J. Parasit. 20:251.

Yakimoff, W.L. and W.F. Gousseff. 1936. Zur Frage der Kokzidien der Füchse. Wien. Tierärztl. Monatschr. 23:359-61.

Yakimoff, W.L. and W.F. Gousseff. 1936a. A propos des coccidies des oiseaux sausages. Ann. Parasit. 14:449-56.

Yakimoff, W.L., P.S. Iwanoff-Gobzem, and S.N. Matschoulsky. 1936. Zur Frage der Infektion der Tiere mit heterogenen Kokzidien. IV. und V. Mitteil. Zentralbl. Bakt. I. Orig. 137:299-302.

Yakimoff, W.L. and N. Kohl-Yalkimoff. 1912. Zur Grage über den Haemoparasitisaus der Seefische. 2-ter Aufsatz. Jarjev. Z. Wiss. Prakt. Vet. Med. 6:1-30.

Yakimoff, W.L. and E.N. Lewkowitsch. 1932. *Isospora theileri* n. sp., Coccidie der Schakale. Arch. Protistenk. 77:533-37.

Yakimoff, W.L. and S.N. Machul'skii. 1940. Koktsidii zhivotnykh soologicheskogo sada v Tashkente. (Coccidia of the animals of the Tashkent zoo.) Parasit. Sborn., Zool. Inst. Akad. Nauk SSSR, Leningrad. 1940(8):236-48. (*Non vidi.*)

Yakimoff, W.L. and I.L. Matikaschwili. 1932. Coccidiosis of skunks. Ann. Trop. Med. Parasit. 25:539–44.

Yakimoff, W.L. and I.L. Matikaschwili. 1933. Die Coccidiose der ussurischen Waschbären. Arch. Protistenk. 81:166–78.

Yakimoff, W.L., I.L. Matikaschwili, and E.F. Rastegaieff. 1933. Zur Frage über die Coccidien der Schakale, Eimeria dutoiti n. sp. Arch. Protistenk. 80:177–78.

Yakimoff, W.L., I.L. Matikaschwili, E.F. Rastegaieff, and E.N. Lewkowitsch. 1933. Coccidia of the Felidae. Parasitology 25: 389–91.

Yakimoff, W.L. and S.N. Matschoulsky. 1935. As coccidioses dos ursos, lobos e cães selvagens. Arch. Inst. Biol. S. Paulo 6:171–77.

Yakimoff, W.L. and I.I. Sokoloff. 1934. Die Sarkozysten des Renntieres und des Maral (Sarcocystis gruneri n. sp.). Berl. Tierärztl. Wochenschr. 50:772–74.

Yakimoff, W.L. and S.K. Terwinsky. 1931. Die Coccidiose des Zobeltieres. Arch. Protistenk. 73:56–59.

Zasukhin, D.N. and N.A. Gaisky. 1930. Toxoplasma nikanorovi n. sp.—novyi kroveparalit stennogo suslika Citellus pygmaeus Pallas. Vestnik Mikrobiol. Epidemiol. Parazitol. 15:27–44.

Figures and Plates

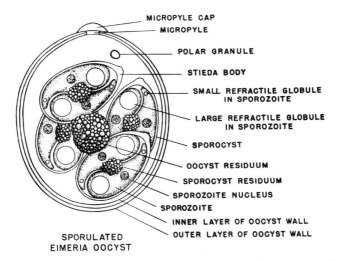

Text Figure 1. Structures of sporulated *Eimeria* oocysts (from Levine, 1973).

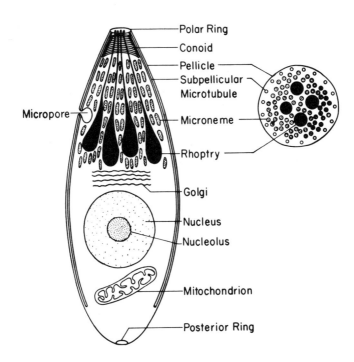

Text Figure 2. Apicomplexan structures (from Levine, 1978).

Text Figure 3. Life cycle of the chicken coccidium *Eimeria tenella.*
A sporozoite (1) enters an intestinal epithelial cell (2), rounds up,
grows, and becomes a first-generation meront (3). This produces a
large number of first-generation merozoites (4), which break out
of the host cell (5), enter new intestinal epithelial cells (6), round
up, grow, and become second-generation meronts (7, 8). These
produce a large number of second-generation merozoites (9, 10)
which break out of the host cell (11). Some enter new host intestinal epithelial cells and round up to become third-generation meronts (12, 13), which produce third-generation merozoites (14).
The third-generation merozoites (15) and the great majority of
the second-generation merozoites enter new host intestinal epithelial cells. Some become microgamonts (16, 17), which each produce
a large number of microgametes (18). Others turn into macrogametes (19, 20). The macrogametes are fertilized by the microgametes
and become zygotes (21), which lay down a heavy wall around
themselves and turn into young oocysts. These break out of the
host cell and pass out in the feces (22). The oocysts then begin to
sporulate. The sporont throws off a polar body and forms four
sporoblasts (23), each of which forms a sporocyst containing two
sporozoites (24). When the sporulated oocyst (24) is ingested by
a chicken, the sporozoites are released (1). (From Levine, 1973).

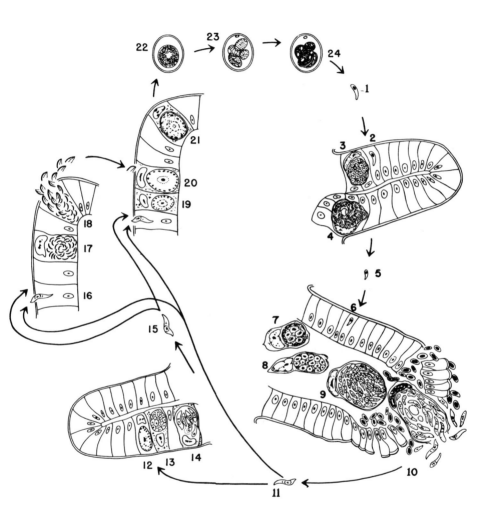

Plate 1

1. *Isospora canis* Nemeséri, 1959 from *Canis familiaris* (from Levine and Ivens, 1965b). 2200X.

2. *Sarcocystis tenella* (?) (Railliet, 1886) Moulé, 1886 from *Canis familiaris* (from Pande, Bhatia, and Chauhan, 1972—cited as *Hoareosporidium pellerdyi*).

3. *Eimeria canis* Wenyon, 1923 from *Canis familiaris* (from Wenyon, 1923). 1500X. Sporulated oocyst without outer wall.

4. *Eimeria canis* Wenyon, 1923 from *Canis familiaris* (from Wenyon, 1923). 1500X. Sporulating oocyst with outer wall intact.

5. *Eimeria canis* Wenyon, 1923 from *Canis familiaris* (from Wenyon, 1923). 1500X. Sporulated oocyst of the smaller size.

6. *Sarcocystis cruzi* (Hasselmann, 1926) Wenyon, 1926 or *S. tenella* (Railliet, 1886) Moulé, 1886 from *Canis familiaris* (from Levine and Ivens, 1965b—cited as a free sporocyst of *Isospora rivolta*). 2200X.

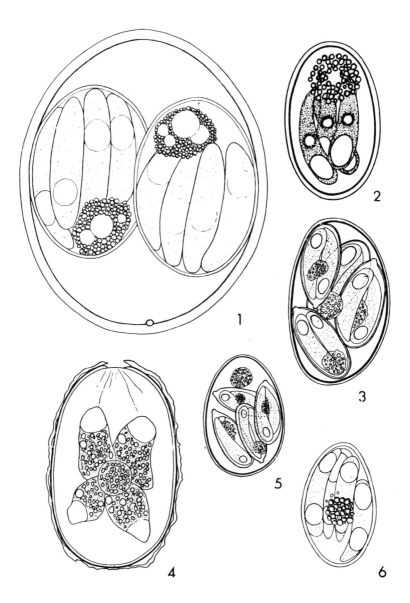

1

2

3

4

5

6

Plate 2

7. *Isospora canis* Nemeséri, 1959 from *Canis familiaris* (from Lepp and Todd, 1974). 1300X.

8. *Eimeria adleri* Yakimoff and Gousseff, 1936 from *Vulpes vulpes* (from Yakimoff and Gousseff, 1936).

9. *Isospora bahiensis* de Moura Costa, 1956 from *Canis familiaris* (from Levine and Ivens, 1965b—cited as *I. bigemina*). 2200X.

10. *Isospora fennechi* Prasad, 1965 from *Fennecus zerda* (from Prasad, 1961). 1100X.

11. *Klossia* sp. Golemansky, 1975 from *Vulpes vulpes* (from Golemansky, 1975). 1000X.

12. *Eimeria vison* Kingscote, 1935 from *Mustela vison* (from Levine, 1948). 2000X.

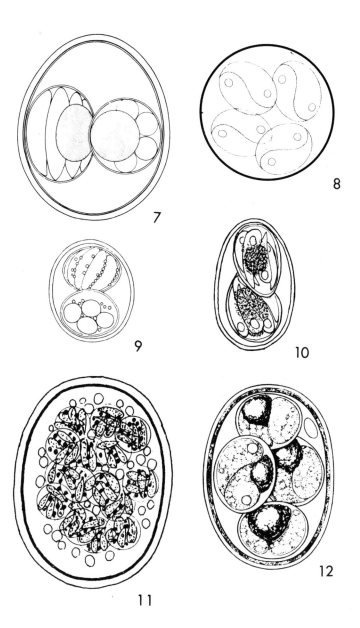

7

8

9

10

11

12

Plate 3

13. *Isospora ohioensis* Dubey, 1975 from *Canis familiaris* (from Levine and Ivens, 1965b—cited as *I. rivolta*). 2200X. Normal sporulated oocyst.

14. *Isospora neorivolta* Dubey and Mahrt, 1978 from *Canis familiaris* (from Mahrt, 1967—cited as *I. rivolta*).

15. *Isospora ohioensis* Dubey, 1975 from *Canis familiaris* (from Levine and Ivens, 1965b—cited as *I. rivolta*). 2200X. Abnormal sporulated oocyst.

16. *Isospora burrowsi* Trayser and Todd, 1978 from *Canis familiaris* (from Trayser and Todd, 1978). 2100X.

17. *Eimeria irara* (Carini and da Fonseca, 1938) from *Eira barbari* (from Carini and Fonseca, 1938). 1400X.

18. *Isospora ohioensis* Dubey, 1975 from *Canis familiaris* (from Levine and Ivens, 1965b—cited as *I. rivolta*). 2200X. Dumbbell-shaped sporulated oocyst.

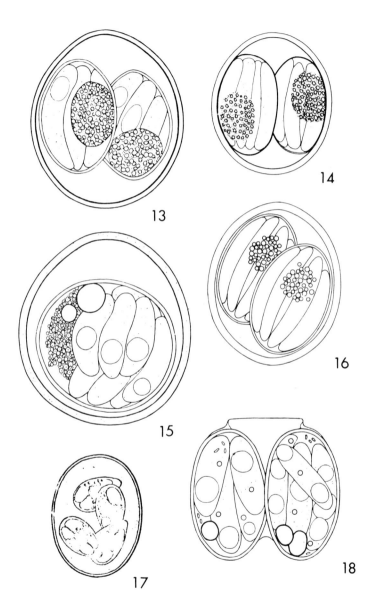

13

14

15

16

17

18

Plate 4

19. *Isospora ohioensis* Dubey, 1975 from *Canis familiaris* (from Dubey, 1975). 3200X.

20. *Eimeria beissini* Svanbaev, 1956 from *Vulpes corsac* (from Svanbaev, 1956). 900X.

21. *Isospora fonsecai* Yakimoff and Machul'skii, 1940 from *Ursus arctos isabellinus* (from Yakimoff and Machul'skii, 1940).

22. *Isospora africana* Prasad, 1961 from *Poecilictis libyca alexandrae* (from Prasad, 1961a). 600X.

23. *Eimeria li* Golemansky, 1975 from *Vulpes vulpes* (from Golemansky, 1975). 1400X.

24. *Eimeria lomarii* Dubey, 1963 from *Vulpes bengalensis* (from Dubey, 1963). 3000X.

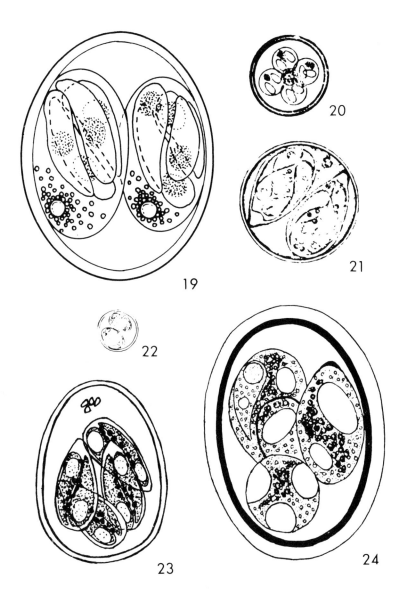

19

20

21

22

23

24

Plate 5

25. *Isospora vulpina* Nieschulz and Bos, 1933 from *Vulpes vulpes* (from Bledsoe, 1976). 1600X.

26. *Isospora vulpina* Nieschulz and Bos, 1933 from *Vulpes vulpes* (from Nieschulz and Bos, 1933). 1500X.

27. *Isospora canivelocis* (Weidman, 1915) Wenyon, 1923 from *Vulpes vulpes* (from Sprehn and Cramer, 1931).

28. *Eimeria nuttalli* Yakimoff and Matikaschwili, 1933 from *Procyon lotor* (from Inabnit, Chobotar, and Ernst, 1972). 3000X.

29. *Isospora canivelocis* (Weidman, 1915) Wenyon, 1923 from *Vulpes vulpes* (from Sprehn and Cramer, 1931).

30. *Isospora canivelocis* (Weidman, 1915) Wenyon, 1923 from *Vulpes vulpes* (from Sprehn and Cramer, 1931).

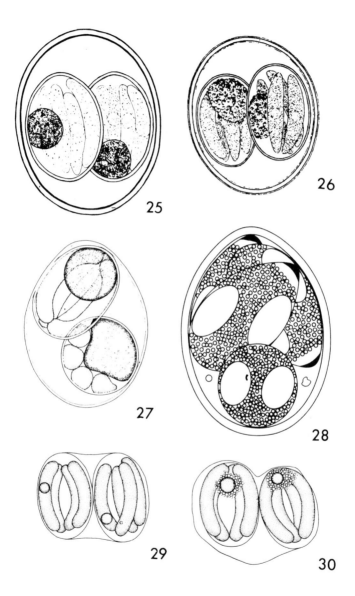

25

26

27

28

29

30

Plate 6

31. *Eimeria borealis* Hair and Mahrt, 1970 from *Euarctos americanus* (from Hair and Mahrt, 1970). 2000X.

32. *Eimeria nasuae* Carini and Grecchi, 1938 from *Nasua nasua* (from Carini and Grecchi, 1938). 2000X.

33. *Eimeria furonis* Hoare, 1927 from *Mustela putorius* var. *furo* (from Hoare, 1927). 2000X. Sporulated oocyst.

34. *Eimeria furonis* Hoare, 1927 from *Mustela putorius* var. *furo* (from Hoare, 1927). 2000X. Oocyst with four sporoblasts.

35. *Eimeria furonis* Hoare, 1927 from *Mustela putorius* var. *furo* (from Hoare, 1927). 2000X. Sporocyst.

36. *Eimeria albertensis* Hair and Mahrt, 1970 from *Euarctos americanus* (from Hair and Mahrt, 1970). 2200X.

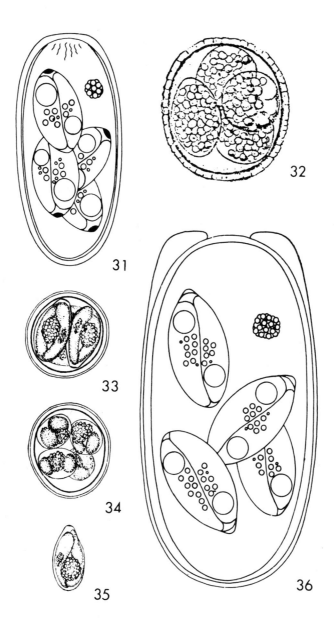

31

32

33

34

35

36

Plate 7

37. *Isospora chobotari* n. sp. from *Procyon lotor* (from Inabnit, Chobotar, and Ernst, 1972—cited as *Isospora* sp.). 3000X.

38. *Eimeria procyonis* Inabnit, Chobotar, and Ernst, 1972 from *Procyon lotor* (from Inabnit, Chobotar and Ernst, 1972). 3000X.

39. *Eimeria mephitidis* Andrews, 1928 from *Mephitis mephitis* (from Andrews, 1928). 2200X.

40. *Isospora zorillae* Prasad, 1961 emend. Pellérdy, 1963 from *Poecilictis libyca alexandrae* (from Prasad, 1961—cited as *I. bigemina* var. *zorillae*). Sporulated oocyst. 900X.

41. *Isospora zorillae* Prasad, 1961 emend. Pellérdy, 1963. Sporulated sporocyst. 900X.

42. *Isospora hoogstraali* Prasad, 1961 from *Poecilictis libyca alexandrae* (from Prasad, 1961). Sporulated oocyst. 500X.

43. *Isospora hoogstraali* Prasad, 1961. Sporulated sporocyst. 500X.

44. *Eimeria hiepei* Gräfner, Graubmann, and Dobbriner, 1967 from *Mustela vison* (from Gräfner, Graubmann, and Dobbriner, 1967). 3750X.

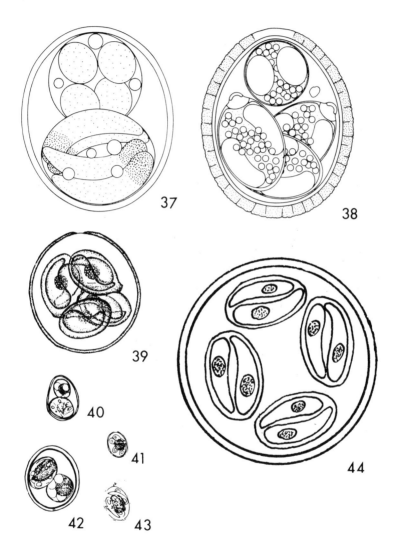

37

38

39

40

41

42   43

44

Plate 8

45. *Eimeria poti* Lainson, 1968 from *Potos flavus* (from Lainson, 1968). 2700X.

46. *Isospora herpestei* Levine, Ivens and Healy, 1975 from *Herpestes auropunctatus.* 1500X.

47. *Isospora sengeri* Levine and Ivens, 1964 from *Spilogale putorius ambarvalis* (from Levine and Ivens, 1964). 2600X.

48. *Isospora hoarei* Bray, 1954 from *Helogale undulata rufula* (from Bray, 1954). 1166X.

49. *Eimeria hiepei* Gräfner, Graubmann, and Dobbriner, 1967 from *Mustela vison* (from Gräfner, Graubmann, and Dobbriner, 1967). 3750X.

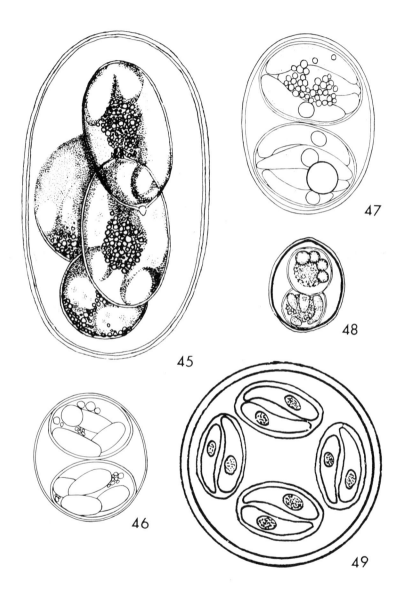

45

47

48

46

49

Plate 9

50. *Isospora laidlawi* Hoare, 1927 from *Mustela putorius* var. *furo* (from Hoare, 1927). Side view. 2000X.

51. *Eimeria anekalensis* Rajesekariah et al., 1971 from *Leo pardus* (from Rajesekariah et al., 1971). 1250X.

52. *Eimeria hammondi* Dubey and Pande, 1963 from *Felis chaus* (from Dubey and Pande, 1963). 1800X.

53. *Eimeria novowenyoni* Rastegaieff, 1929 from *Leo pardus* (?) (from Rajesekariah et al., 1971). 1100X.

54. *Isospora laidlawi* Hoare, 1927 from *Mustela putorius* var. *furo* (from Hoare, 1927). End view. 2000X.

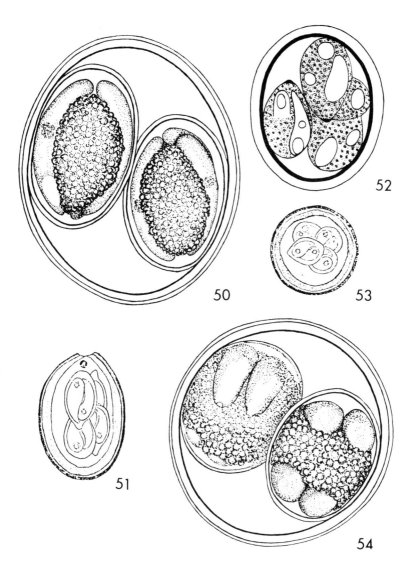

50

52

53

51

54

Plate 10

55. *Isospora buriatica* Yakimoff and Machul'skii, 1940 from *Vulpes bengalensis* (from Dubey, 1963). 2200X.

56. *Eimeria ictidea* Hoare, 1927 from *Mustela putorius* var. *furo* (from Hoare, 1927). 2000X.

57. *Eimeria mustelae* Iwanoff-Gobzem, 1934 from *Mustela vison* (from Levine, 1948). 2100X.

58. *Sarcocystis putorii* (?) (Tadros and Laarman, 1976) nov. comb. from *Mustela vison* (from Levine, 1948—cited as *Isospora bigemina*). 2100X.

59. *Isospora laidlawi* Hoare, 1927 from *Mustela vison* (from Levine, 1948). 4200X.

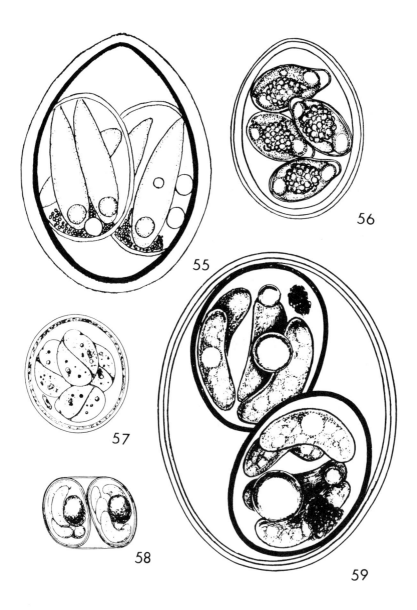

55

56

57

58

59

Plate 11

60. *Isospora mungoi* Levine, Ivens, and Healy, 1975 from *Herpestes edwardsi* (from Dubey and Pande, 1963—cited as *I. garnhami*). 2300X.

61. *Eimeria felina* Nieschulz, 1924 from *Felis catus* (from Nieschulz, 1924). 2000X.

62. *Isospora garnhami* Bray, 1954 from *Helogale undulata rufula* (from Bray, 1954). 1166X.

63. *Isospora dasguptai* Levine, Ivens, and Healy, 1975 from *Herpestes edwardsi* (from Dubey and Pande, 1963—cited as *I. garnhami*).

64. *Eimeria felina* Nieschulz, 1924 from *Felis chaus* (from Dubey and Pande, 1963). 2500X.

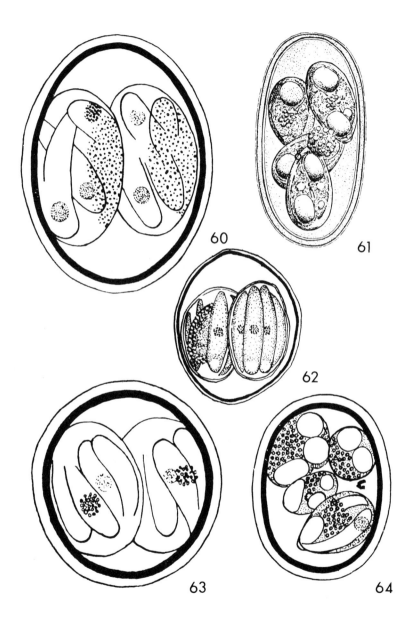

60

61

62

63

64

Plate 12

65. *Isospora rivolta* (Grassi, 1879) Wenyon, 1923 from *Felis catus* (from Shah, 1970). 2200X.

66. *Eimeria mathurai* Dubey and Pande, 1963 from *Felis chaus* (from Dubey and Pande, 1963). 2100X.

67. *Eimeria mathurai* Dubey and Pande, 1963 from *Felis chaus* (from Dubey and Pande, 1963). 2200X.

68. *Eimeria* (?) *hartmanni* Rastegaieff, 1930 from *Leo pardus* (?) from Rajesekariah et al., 1971). 1300X.

69. *Isospora felis* Wenyon, 1923 from *Felis catus* (from Shah, 1970). 2100X.

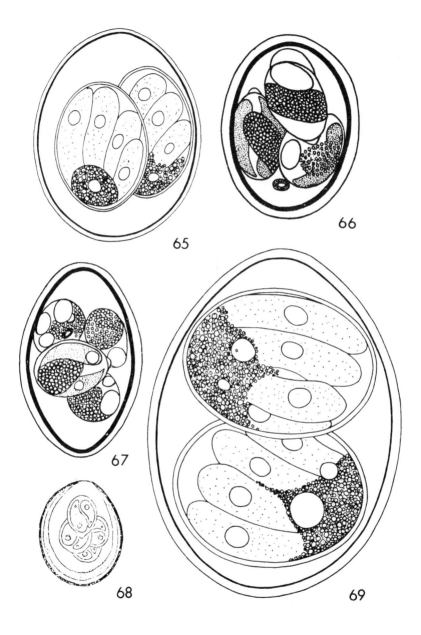

65

66

67

68

69

Plate 13

70. *Eimeria newalai* Dubey and Pande, 1963 from *Herpestes mungo* (from Dubey and Pande, 1963). 3600X.

71. *Eimeria newalai* Dubey and Pande, 1963. Sporulated sporocyst. 3800X.

72. *Isospora melis* Pellérdy, 1965 from *Meles meles* (from Pellérdy, 1965). 2000X.

73. *Isospora pellerdyi* Dubey and Pande, 1963 from *Herpestes mungo* (from Dubey and Pande, 1963). 2200X.

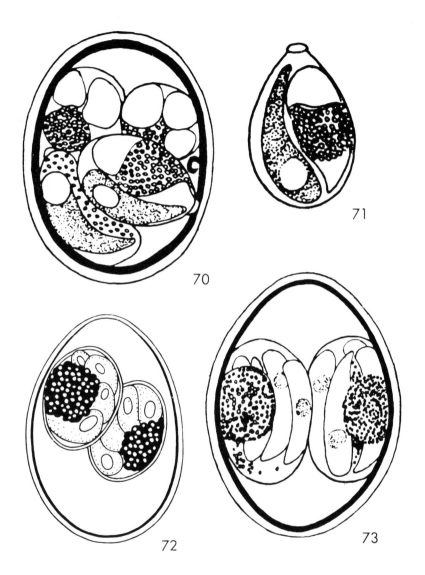

70

71

72

73

Plate 14

74. *Isospora spilogales* Levine and Ivens, 1964 from *Spilogale putorius ambarvalis* (from Levine and Ivens, 1964). 2600X.

75. *Isospora viverrae* Adler, 1924 from *Civettictis civetta* (from Adler, 1924). 1250X.

76. *Eimeria cati* Yakimoff, 1933 from *Felis chaus* (from Dubey and Pande, 1963). 2500X.

77. *Isospora levinei* Dubey, 1963 from *Hyaena hyaena* (from Dubey, 1963). 2100X.

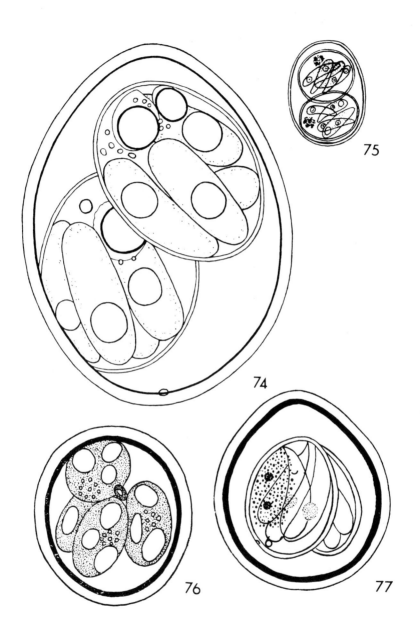

74

75

76

77

Plate 15

78. *Eimeria cati* Yakimoff, 1933 from *Felis catus* (from Yakimoff, 1933).

79. and 80. *Eimeria ursi* Yakimoff and Matschoulsky, 1935 from *Ursus arctos* (from Yakimoff and Matschoulsky, 1935).

81. *Isospora dutoiti* Yakimoff, Matikaschwili, and Rastegaieff, 1933 from *Canis aureus* (from Yakimoff, Matikaschwili and Rastegaieff, 1933).

82. *Isospora eversmanni* Svanbaev, 1956 from *Mustela eversmanni* (from Svanbaev, 1956). 900X.

83. *Isospora theileri* Yakimoff and Lewkowitsch, 1932 from *Canis aureus* (from Yakimoff and Lewowitsch, 1932.

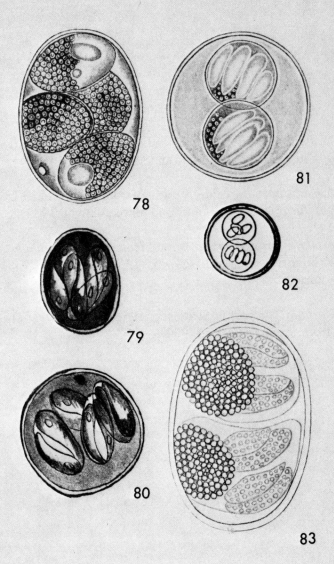

78

81

79

82

80

83

Plate 16

84. *Eimeria nuttalli* Yakimoff and Matikaschwili, 1933 from *Procyon lotor* (from Yakimoff and Matikaschwili, 1933).

85. *Eimeria nuttalli* Yakimoff and Matikaschwili, 1933 from *Procyon lotor* (from Yakimoff and Matikaschwili, 1933).

86. *Eimeria voronezhensis* n. sp. from *Mephitis mephitis* (from Yakimoff and Matikaschwili, 1932—cited as *E. mephitidis* Andrews, 1928).

87. *Isospora pavlovskyi* Svanbaev, 1956 from *Mustela eversmanni* (from Svanbaev, 1956). 900X.

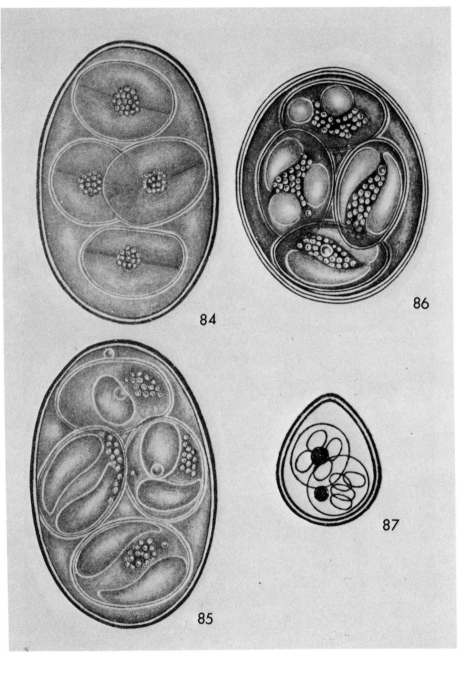

84

86

85

87

Index

HOSTS

A Note on the Authors

Norman D. Levine is professor of veterinary parasitology and veterinary research at the University of Illinois. He received his B.S. from Iowa State University in 1933 and his Ph.D. from the University of California at Berkeley in 1937. He has served as president of the Society of Protozoologists (1959–60), president of the American Society of Professional Biologists (1967–69), president of the Illinois State Academy of Science (1966–67), president of the American Microscopical Society (1968–69), and is now president of the American Society of Parasitologists (1980), and chairman of the Tropical Medicine and Parasitology study section of the National Institutes of Health (1965–69). He was editor of the *Journal of Protozoology* from 1966 to 1972. His numerous publications include *Protozoan Parasites of Domestic Animals and of Man* (2d ed., 1973), *Preventive Medicine in World War II*, Volume VI (co-author with twelve others, 1963), *Malaria in the Interior Valley of North America: A Facsimile Selection from Daniel Drake's (1850) Book* (1965), *The Coccidian Parasites (Protozoa, Sporozoa) of Rodents* (with Virginia Ivens, 1965), *The Coccidian Parasites (Protozoa, Sporozoa) of Ruminants* (with Virginia Ivens, 1970), *Natural Nidality of Transmissible Diseases with Special Reference to the Landscape Epidemiology of Zooanthroponoses*, by E.N. Pavlovsky, translated by E.K. Plous, Jr. (editor, 1966), *Nematode Parasites of Domestic Animals and of Man* (2d ed., 1980), *Textbook of Veterinary Parasitology* (1979), and *Principal Parasites of Domestic Animals in the United States* (with Virginia Ivens and Daniel L. Mark, 1978).

A 1950 graduate of the University of Illinois, Virginia Ivens is associate professor at the College of Veterinary Medicine at the university. She has co-authored numerous articles in scientific journals and has translated several articles from Russian. She co-authored with Norman D. Levine *The Coccidian Parasites (Protozoa, Sporozoa) of Rodents* (1965) and *The Coccidian Parasites (Protozoa, Sporozoa) of Ruminants* (1970) and was senior author (with Daniel L. Mark and Norman D. Levine) of *Principal Parasites of Domestic Animals in the United States* (1978).